Palgrave Gothic

Series Editor
Clive Bloom
Middlesex University
London, UK

This series of Gothic books is the first to treat the genre in its many interrelated, global and 'extended' cultural aspects to show how the taste for the medieval and the sublime gave rise to a perverse taste for terror and horror and how that taste became not only international (with a huge fan base in places such as South Korea and Japan) but also the sensibility of the modern age, changing our attitudes to such diverse areas as the nature of the artist, the meaning of drug abuse and the concept of the self. The series is accessible but scholarly, with referencing kept to a minimum and theory contextualised where possible. All the books are readable by an intelligent student or a knowledgeable general reader interested in the subject.

More information about this series at
http://www.palgrave.com/gp/series/14698

Holly-Gale Millette · Ruth Heholt
Editors

The New Urban Gothic

Global Gothic in the Age of the Anthropocene

palgrave
macmillan

Editors
Holly-Gale Millette
Winchester School of Art
University of Southampton
Winchester, UK

Ruth Heholt
School of Writing and Journalism
Falmouth University
Cornwall, UK

ISSN 2634-6214 ISSN 2634-6222 (electronic)
Palgrave Gothic
ISBN 978-3-030-43776-3 ISBN 978-3-030-43777-0 (eBook)
https://doi.org/10.1007/978-3-030-43777-0

Cover image: Doug Armand/Alamy Stock Photo
Cover design by eStudioCalamar

This Palgrave Macmillan imprint is published by the registered company Springer Nature Switzerland AG
The registered company address is: Gewerbestrasse 11, 6330 Cham, Switzerland

Contents

Editors and Contributors

About the Editors

Dr. Holly-Gale Millette is a Senior Teaching Fellow at Southampton University UK. A Social and Cultural Historian publishing regularly in international journals and books, recent work has focused on spatial, political and psychosocial representations of urban cultures and also the new Urban Gothic in cultural theory and media texts.

Dr. Ruth Heholt is Senior Lecturer in English at Falmouth University. She is co-editor of several collections including: *Gothic Britain: Dark Places in the Provinces and Margins of the British Isles* (2018) with William Hughes, *The Victorian Male Body* (2018) with Joanne Ella Parsons and *Haunted Landscapes* (2017) with Niamh Downing. She has organised several symposia and is editor of the peer-reviewed journal *Revenant: Critical and Creative Studies of the Supernatural*. She is currently working on a monograph on the Victorian author Catherine Crowe.

Contributors

Dr. Xavier Aldana Reyes is Senior Lecturer in English Literature and Film at Manchester Metropolitan University and a founder member of the Manchester Centre for Gothic Studies. His books include *Spanish Gothic* (2017), *Horror: A Literary History* (editor, 2016), *Horror Film* *and Affect* (2016), and *Body Gothic* (2014). Xavier is chief editor of the Horror Studies book series published by the University of Wales Press.

Dr. Katarzyna Ancuta is a lecturer at Chulalongkorn University in Bangkok, Thailand. Her research interests oscillate around the interdisciplinary contexts of contemporary Gothic/Horror, currently with a strong Asian focus. Her recent publications include contributions to *A New Companion to the Gothic* (2012), *Globalgothic* (2013), *The Cambridge Companion to the Modern Gothic* (2014), *Neoliberal Gothic* (2017), *The Routledge Handbook to the Ghost Story* (2017), and *B-Movie Gothic* (2018). She also co-edited three special journal issues on *Thai* (2014) and *Southeast Asian* (2015) horror film and *Tropical Gothic* (2019), and a collection *Thai Cinema: The Complete Guide* (2018).

Dr. Karl Bell is a Reader in Cultural and Social History at the University of Portsmouth, and Director of the Supernatural Cities project (supernaturalcities.co.uk). He is the author of two monographs, *The Magical Imagination: Magic and Modernity in Urban England, 1780–1914* (2012), and the award-winning *The Legend of Spring-Heeled Jack: Victorian Urban Folklore and Popular Cultures* (2017). His research explores the supernatural and the fantastical imagination from the late-eighteenth to the twentieth century.

Dr. R. M. Francis is a writer from the Black Country. He graduated with distinction from Teesside University's Creative Writing MA and recently completed his Ph.D. at the University of Wolverhampton. He is author of five poetry chapbooks. In 2020 Smokestack Books will publish his first full collection and Wild Pressed Books are publishing his debut novel. He was the inaugural David Bradshaw Writer in Residence 2019 at Oxford University.

Dr. Michael Fuchs is a fixed-term Assistant Professor in American studies at the University of Graz in Austria. He has co-edited six collections, most recently *Intermedia Games—Games Inter Media: Video Games and Intermediality* (2019). In addition, he has authored and co-authored more than a dozen journal articles, which have appeared in venues such as *The Journal of Popular Culture*, *The Journal of Popular Television* and the *Quarterly of Film and Video*, and more than thirty book chapters in volumes such as *The Cambridge History of Science Fiction* (2019), *Ecogothic in Nineteenth-Century American Literature* (2018), *Horror*

Television in the Age of Consumption: Binging on Fear (2018), and *B-Movie Gothic: International Perspectives* (2018). Among others, he is currently putting the finishing touches to a co-edited volume on American urban spaces in the fantastic imagination and working on a monograph on urban spaces in American horror movies.

Dr. Madelon Hoedt is a Senior Lecturer at the Department of Drama, Theatre, and Performance at the University of Huddersfield (UK). Her Ph.D. "Acting Out: The Pleasures of Performance Horror" focuses on genre, performance, stagecraft, and audience affect. In her other research, she is concerned with issues of narrative and embodied experience in live performance and video games (specifically in relation to horror and the Gothic). She is currently working on two monographs, one on immersive horror performance experiences, and one on the Gothic videogame *Bloodborne* (FromSoftware, 2015).

Prof. Tanya Krzywinska is Professor of Digital Games and Director of the Games Academy at Falmouth University. Beginning her academic career when videogames were largely invisible and far from respectable, she's sought to argue for the importance of games as a new art form. She is the author of several books and many articles on different aspects of video games and representations of the Gothic. Currently she is working on a monograph, *Gothic Games*, and, when time allows, she is an artist.

Dr. Annemarie Lopez is an arts writer and the former editor of The Week, Australia. She recently completed a Ph.D. in crime fiction and psychogeography at Macquarie University, Sydney.

Dr. Garth Sabo received his doctorate from the Department of English at Michigan State University. His research draws on material ecocriticism, narratives of science, and scatological theory to explore how waste, particularly human excrement and urine, permeates the cultural imaginary of contemporary English-language fiction. His monograph project tracks cultural relationships to human waste across the twentieth and twenty-first centuries in light of key changes to the scientific, ecological, and aesthetic understanding of bodily waste and the human microbiome.

Dr. Molly Slavin is a Marion L. Brittain Postdoctoral Fellow at the Georgia Institute of Technology (Atlanta, GA, USA), specialising in representations of crime in postcolonial and global Anglophone fiction. She has published articles in the *Journal of Commonwealth Literature*, *Global*

South, Journal of the Midwest Modern Language Association, and *C21 Literature: Journal of 21st Century Writings*.

Leila Taylor completed her Masters in Fine Art at Yale University and an M.A. in Liberal Studies at the New School for Social Research in New York. Her focus was on how the gothic sensibility is interpreted and transformed through the African American experience, mourning as a public practice and the aesthetics of romanticised melancholy.

Dr. Kwasu David Tembo is a Ph.D. graduate from the University of Edinburgh's Language, Literatures, and Cultures department. His research interests include—but are not limited to—comics studies, literary theory and criticism, philosophy, particularly the so-called "prophets of extremity"—Nietzsche, Heidegger, Foucault, and Derrida. He also writes poetry and makes experimental electronic music.

Prof. Gina Wisker is Professor of contemporary literature and higher education at the University of Brighton. She teaches and publishes on Gothic fictions including *Contemporary women's Gothic fiction* (Palgrave Macmillan, 2016), on Margaret Atwood, Angela Carter, Toni Morrison, Nalo Hopkinson, and numerous articles on women's vampire writing, and postcolonial women's Gothic. Gina went to school in Singapore.

Prof. Julian Wolfreys is author and editor of more than forty books on nineteenth and twentieth-century literature. He is also a poet and musician, and has published one novel (so far), *Silent Music* (Triarchy Press).

List of Figures

'Everything Is True': Urban Gothic Meets the Chthulucene in Multiplayer Online Game, *The Secret World*

Rust Belt Ruins: The Gothic Genius Loci of Detroit

Communal After-Living: Asian Ghosts and the City

Suzhou River: 'On the [Haunted] Waterfront'

The New Urban Gothic: Introduction

Holly-Gale Millette

The global politics of the past few years has been dark. With right-wing populism sweeping the north, prejudice, intolerance and violence seem to be relentlessly on the rise. Movements such as *#MeToo* and *Extinction Rebellion* point to the collapse and the sheer unworkability of patriarchy and capitalism, while the old order clings on; increasingly intransigent. In the humanities globally, there is widespread recognition and agreement that philosophic human anthropic time—in its broadest undifferentiated sense—is, and has consistently been, underwritten by the capitalist system that the global economic systems of the neoliberal period have exploited our monstrous desires, power, greed and inequalities to such a degree as to have wreaked unrepairable and irreversible damage to our planet. Increasingly, fear and anxiety are now linked to geo-biological and epidemiological crises.

Perhaps unsurprisingly, much of the focus of the discourse of doom, decay and despair is centred on global urban spaces—spaces that are the foci of the aesthetic politics[1] of hyper-consumption, contagion and marketing of late-capitalism. Today, aesthetic debates from the banal

H.-G. Millette (✉)
Winchester School of Art, University of Southampton, Winchester, UK
e-mail: H.Millette@soton.ac.uk

H.-G. Millette and R. Heholt (eds.),
The New Urban Gothic, Palgrave Gothic,
https://doi.org/10.1007/978-3-030-43777-0_1

to the theoretical have become a proxy for popular politics. Personal intercessions such as *#MeToo* and popular movements such as *Extinction Rebellion*, capture public attention and opinion, while replicating the very class-based entitlement that marginalise some humans over others in the first instance. Similar to its predecessor in *Occupy*, the real darkness in these movements is the precarious and paperless individuals it dis-includes, disaffects and displaces at the end of the public trial or the Oxford Street Parade.

The isolation caused by digital globalisation, the fear and anxiety we have over our geo-biological crises, and the surveillance that geomedia inflicts on us, marginalise us in similar ways. Geomedia and the City is an emerging concept that has been deployed to capture a particular technological condition, associated with recent rapid developments in digital technology that are tied to the dialectics of the locative in surveillance, GPRS and 'smart cities'. However, the concept of geomedia carries deeper/wider ontological and epistemological concerns that transcend the simple twining of geography and media. Geomedia has the ability to entice, enable, regulate, constrain or prohibit access to public space. Increasingly, we are seeing enclaves, zones, slums, camps, gated communities and other spaces of exception developing as the loci of our millennial gothic protagonists and we are encountering gothic representations of surveillance and affect that contribute to our mental imbalance and decay. These mediations over the flows and breaks in urban life, what constitutes inclusion or exclusion—the right to the city (Lefebvre 1991/1974)—residue and hauntings, and precarious urbanities are all finding their way into Gothic fictions.

Precarity and marginalisation has, historically, been easiest to perceive in urban environments and today's urban landscapes have been so soaked with inequality and toxicity as to become 'anti-landscapes'. The term anti-landscapes refers to loci that have been so degraded as to become pervasively toxic in real terms. Flint, Michigan is an example, where, in 2014, local government corruption exacerbated already poverty-stricken residents with a drinking water crisis that has yet to be resolved and left residents without potable drinking water on top of one of the worst unemployment rates in the country. Inhabitants respond to such places with dread, foreboding and aversion. Nye and Elkind believe that the 'concept of the anti-landscape provides a point of unity for work that deals with sustainability, the limits to growth, dystopia, environmental degradation, "the tragedy of the commons," and the overuse of resources'

(2014, p. 14) and this book is underwritten by these themes and by the modernist wastelands of excess, hubris and moral decay that now typify our cities. Landscape has always been an integral part of Gothic, horror and crime fictions as it is in the urban that we can see the human imprint most starkly. As deployed in fictions, the new urban Gothic landscape is almost never an exhaustive description of a place, but rather a series of references that suggest the feel or surrender of a place to the reader—modified urban areas where people live a collective existence of, all too often, disaffectation (Jackson 1984, p. 8). As Lee Horsley reminds us,

> the cityscape … the metaphoric dark alleyway and the associated geography of the labyrinthine big city are not just a tidy source of clues. Much less comprehensible than Holmes's London, this is an intractable, uncontainable, ultimately unknowable terrain, to be grasped only in a fragmentary way. (2005, p. 71)

While *The New Urban Gothic* attempts a move away from the iconic Gothic cities of London and New York, the city spaces in the essays in this collection are also 'intractable, uncontainable' (Warwick 1999, p. 251) and unruly. Threatening yet thrilling, these sublime cities are, as Warwick continues, 'seen as uncanny, constructed by people yet unknowable by the individual'. Thus, and to some degree, the new urban Gothic is a tale of survival against the very darkness of the enclosing anti-landscape. To survive in them, people must find ways to live within and maintain their landscapes, lest they become spaces that are unable to sustain life. Anti-landscapes of the new urban Gothic (and anxieties over anti-landscapes in the new urban Gothic), are therefore post-human in that they suggest—whether temporary or long term—that human beings have created environments in which their communities may (and very soon) become unsustainable for their species. Post-humanism, as it is used here, is a discourse that is characterised by a relationship between humans and their hybrid and/or non-human others, thus exhibiting a tendency to think beyond humanism, anthropomorphism or anthropocentrism.

Environments—specifically urban ones—that have become so polluted they represent a toxic threat (physical and psychical) to humanity, have featured in popular fiction for generations. Now, in the millennium, ideas of toxicity have broader, more intersectional, definitions that reach beyond traditional references to physical or chemical residuum. Millennial toxicity is both concrete and atmospheric, for example: masculinities can be toxic, as can cultures of work or asset-structuring. Identifying phenomena as toxic and monstrous, is something this volume does not shy away from discussing. When and where practices, attitudes and structures are constituted as toxic and monstrous, humanity is de-centred and vulnerable. Pertinent to this volume, is the understanding that globally inequality has created cultures of toxicity that are enacting a 'slow violence' (Nixon 2011) on humanity. Expedited fetishistic deterioration via what Jason Moore (2015) calls 'the Four Cheaps' (cheap: petroleum-based energy, raw materials and minerals, modified food and concentrated off-shored labour) are dark Gothic violations that result in differentiated polarities of experience, suffering and demands for justice globally. The awareness of such violence and contamination is higher in urban areas where renewed debates about freedom, speech practices, trigger warnings and safe spaces proliferate. Such toxicity reproduces attitudes of (and to) Gothic identity and history—identities and histories as discussed in the first part of this collection. The meanings, functions, and residue of such toxicity – how toxicity is produced, sustained, distributed and left as waste—has become a central narrative in our new urban Gothic narratives and this collection begins with a reflection on that.

The Gothic tropes found in many of our new urban storyworlds are, or act as, bridges between aesthetics politics, popular ideology and psychosocial haunting. This book argues that such a hybrid reinterpretation of the Gothic has emerged and is here to stay. Far from 'flattening out' the Gothic in these narratives, Jameson argues (2013) that such an interdisciplinary approach adds dimension, positionality and depth to the darker storyworlds and fictions of the new urban Gothic. In this way, *The New Urban Gothic: Global Gothic in the Age of the Anthropocene* examines representations of the Gothic across many different media forms to explore the global dystopic and grotesque imaginings of the darker side of contemporary life and it uses the Anthropocene as a sociocultural critique of our anxieties of living in it. It does this from four significant entrypoints: (1) psychosocial affect; (2) material itself as transmedia; (3) intersectionalities in its narratives' subjects and (4) the various global responses to death, trauma and/or the end of nature. These essays mark how our bleak tales have taken on a new lease of life by being transmediated in new digitised, mediatised and socialised forms. By transmedia storytelling we mean narratives that 'represent a process where integral elements of a fiction get dispersed systematically across multiple delivery channels for the purpose of creating a unified and coordinated entertainment experience' (Jenkins 2009, p. 56). Texts that form this collection are streamed and Tweeted, shared on multiplayer gaming platforms, and are incorporated and developed by fan fiction response. Their storyworlds are digital, visual, fashioned as games, gone in a second and electrified. They rest on graveyards of debt (Detroit); they burrow underground (Moscow) in the Metro; they rupture from wells of haunted regeneration (The Black Country); they marketise with tropes of revenant nationalism (Barcelona) and they linger as colonial ghosts in the streets (Shanghai), in high-rise apartments (Seoul, Hong Kong) and in cemeteries (Delhi). They make scarce that which should be readily available and spread contagions of loneliness (as in the Anime, *Angel's Egg*). Choosing to explore the aesthetic political configurations of Gothic in the age of the Anthropocene invites the kind of critique that treats this material and its audience both stylistically and affectively. The value of this ideological formation to Gothic Studies is that it allows a greater breadth and depth to be revealed in the Gothic tropes used in late twentieth and early twenty-first-century texts. Which is to say that it is employed here specifically to connect global meta-histories of the systems of domination, periphery, anti-landscape and residue.

The Anthropocene is our current lived epoch. It has been the subject of increasing attention since its acceptance by the earth sciences (GSA 2011, 9th of October—'Presidential Address') in 2011. Since then it has been the focus of a great deal of scholarship in the humanities, cultural studies, art, digital media and literary studies. The modern era's digital revolution has, arguably, increased popular awareness and anxieties over the Anthropocene's key tenets and its queries about ecology, contagion, climate change, the nuclear age, poverty, hunger and the irrevocable and dystopian effect that humans have had on our planet. In fact it might be more accurate to say that the current era is that which *recognises* the Anthropocene, as it has been with us for generations. A central argument is that today's Gothic is always already post-human and that it is only the effect of our heretical hubris that is now being recognised as monstrous. *The New Urban Gothic*, looks to the urban areas which concentrate the residue of human habitation's affect on the planet and showcase the decay, filth, overcrowding and cruelty of the anti-landscape to explore this effect. But discussing Gothic Studies this way places Gothic landscapes—anti or otherwise—within a linear progression of epistemes that, to some degree, have become collapsed and confused. Indeed, a significant contention at the 2018 Conference of the International Gothic Studies Association was how—or if—we should discuss the Anthropocene in Gothic Studies. If the form and function of Gothic temporalities in the millennium no longer holds to the palimpsestic character that Walter Benjamin (1969) identified in it, nor the liquidity Bauman (2012) diagnosed in late modernity, or the critical reflexivity that Ulrich Beck (1994) also pointed to in the late twentieth century, on what terms should we discuss Gothic in the Age of the Anthropocene? There is a general agreement here that conceiving time in Gothic Studies as being merely deterministic or material perspectives of space (as signalled by the 'spatial turn': Lefebvre 1991/1974; Massey 1994; Harvey 1996), or tightened, static paradigms of phenomena such as the frontiers or communities that emerged out of the 'identic turn' in the late 1990s (Agamben 1993; Nancy 2000), fails to allow us to trace the phenomenological aspects of the Gothic in the millennium with any degree of elasticity and intersectionality—something this volume was intent on. So, and following Crang (1998)—a cultural geographer—this volume conceives of *The New Urban Gothic in the Age of the Anthropocene* as an expansive and unstable phenomenon, specific to the place-time of the millennium—what is discussed in the introduction to the second part of this collection as 'diachronicity'. The various chapters presented here show our sociopolitical fabric as unstable and fragmenting. They show human-kind as having undergone—or on the brink of undergoing—a metamorphosis at a deep ethical level that is specific to the global states of an emergency our species finds itself in. Millenial activism and aesthetic practices have moved us well beyond the high/low-brow distinctions of Bourdieu and this new turn has caused us to reassess the validity and utility of the hubris of the human that influenced the rise of Gothic Studies in the mid-eighteenth century. Contemporary notions of Gothic is increasingly distancing itself from the medieval and romanticised notions of the human as the first set out in the Promethean philosophy on which Shelley wrote *Frankenstein* (1818). The white, male, land-owning human no longer takes precedence as the Cartesian subject and the Monster is no longer external and alien to it. Today's Gothic is post-human; it eschews the *Frankenstein* referential as a heretical positioning of the human. Evolution, philosophy and higher mathematics have provided a new essential viewpoint on what it is to be human—one that condemns such narrow homocentric thinking. In a universe that is so patently plan-less, it is heresy to believe that humanity is at the centre of it; that humans are central to the larger plan of creation. If proof is needed, just look at what we have done to the planet, which, for centuries, we have harnessed and attempted to control from the centralising Cartesian perspective. Such arrogance has cost us everything.

The world, it seems is literally falling apart and the fictions and representations of the twenty-first century reflect this. Robert Mighall suggests that: 'The "anthropological" focus of late-Victorian Gothic involved a double movement: outwards to the margins of the Empire, and inwards to focus on the domestic "savages" which resided in the very heart of the civilized world' (2007, p. 136). The third and final part of this book explores these monsters 'within', crouching in the pools of darkness that are cast by the lights of cities around the globe. Here what were margins of the Empire are now cities and anti-landscapes that are central to the crisis. *The New Urban Gothic* examines these global fictions of falling apart and explores our preoccupation with social inequalities, precarity and the peripheral. Our new urban Gothic is made up of these repeated stories; of these characters and character-types that resurface and continue to haunt the contemporary imagination. As Carol Clover maintains in relation to cheap horror movies, the urban Gothic is a genre which celebrates 'the free exchange of themes and motifs, the archetypal characters and situations, the accumulation of sequels, remakes, imitations. This is a field in which there is in some sense no original, no real or right text, but only variants' (1992, p. 11). Because of this, these new fictions refer to and often remediate more traditional genres and stories where dark tales of urban life are told, through literature, film and televisual texts. This essay collection explores the resurgence of serialised grotesque and retold narratives of degeneration, ecological and economic ruin, dystopia, mechanised future inequality and crime in recent fictions set in the urban space. *The New Urban Gothic* looks to the intersections of time, space and media when examining the different and shifting meanings that arise from contemporary texts that draw on the tradition of the Victorian urban Gothic in their storytelling.

Often dystopic, the urban Gothic was pioneered in the mid-nineteenth century in Britain and the United States and developed in serialisations such as R. L. Stevenson's *The Strange Case of Dr. Jekyll and Mr. Hyde* (1886); novels such as Oscar Wilde's *The Picture of Dorian Gray* (1890); novellas such as Stephen Crane's *Maggie: A Girl of the Streets* (1893) and dramaturgy including Gaston Leroux's *The Phantom of the Opera* (1909). Much has been written since on nineteenth-century urban Gothic fiction but less has been written on millenial reimaginings of the urban Gothic. In the late twentieth century, there seemed to be a perspectival backward step made in the revival of urban Gothic tropes, notably with authors of Vampire fiction such as Anne Rice, and graphic novels that drew on dark city landscapes of the nineteenth-century topos. However, and since then, the genre has expanded in the serialised narratives that have proliferated in the on-demand streaming networks of Netflix, Amazon Prime and others. These new fictions have signalled a new era for the urban Gothic and the anxieties over its crime, corruption, child and sexual abuse, supernatural hauntings, serial killers, dystopic science fiction and Worlds End narratives. Responding to this upsurge, this book offers an analysis and discussion of the cultures of at least one global city in the millennium as marked by the Gothic. Plots in the *New Urban Gothic* are permeable to the strangeness of what is new in the millennium. Mindful of the new space/time that is created by digital media and virtual worlds and of the hegemonic tendency of globalisation, the Gothicism we are seeing in contemporary narratives question and deconstruct the Gothic in new and ever more global ways.

Off-screen and in our streets, the discourse of the climate change emergency and the pressure group *Extinction Rebellion* (among others), focuses our anxieties on our species' imminent eradication at the hands of a disintegrating natural environment. As rebellion.earth, the official website of *Extinction Rebellion* states:

> The air we breathe, the water we drink, the earth we plant in, the food we eat, the beauty and diversity of nature that nourishes our psychological well-being – our very health – are all being corrupted and compromised by the human values behind our political and economic systems and consumer-focussed lifestyles. (https://rebellion.earth/the-truth/the-emergency/)

The World Health Organisation estimates 4.2 million deaths a year from ambient air pollution, most of which will be emanating from cities around the world (https://www.who.int/airpollution/ambient/en/). To say nothing of the rapidly mutating and drug-resistant viruses that will almost certainly kill us all before the earth falls apart. And the seat of this human-centred 'corruption' must be the metropolises—the urban centres where people are crammed together and pollute the air to the extent that it is sometimes difficult to see, let alone to breathe. Our cities are the source of shelter and obfuscation for the destructive forces and systems that fuel our self-extinction and annihilation. Every four-year-old child that engages with a game of 'Top Trumps – Predators', learns that the most dangerous and successful predator on earth is the animal: Homosapien. Death, globally, has consistently been at the hands of devaluing and destroying other humans and other species for greed's sake. Sara Wasson says that as 'the location of menace changes, so does the nature of the threat: while earlier Gothic generates a thrill of fear which ghosts or demons, the metropolis evokes the horrors of human violence and corruption' (2010, p. 3). This corruption has always been concentrated in—and seeps out from—the cities. The demons in the *New Urban Gothic* are the residue of this leaching. As Warwick noted in 1999, 'the sense of the incipient destruction of the physical environment is a strong and characteristic Gothic device' (p. 79) and we are now clear, that this 'incipient destruction' is the devastation that humans have wrought to their epoch, that its spaces and anxieties are Gothic and concentrated in our cities, and that it will see the end of our current lived era—the Anthropocene.

Contents

'So what makes the urban Gothic?' asks Mighall (2007, p. 57). His answer, in part, is that it needs to be '*of* a city rather than just in a city' (2007, p. 57). The city is not just a location onto which the action is imposed, the Gothic permeates the streets and the buildings, the stinking rubbish heaps and the sulphurous and poisonous rivers. Yet, inevitably, there is not one single unified idea of the urban Gothic. As in older Gothic fiction, different locales affect the tales of the urban and imbue them with a specific sense of place and space. As Roger Luckhurst suggests: 'the ghosts of London are different from those of Paris, or those of California' (2002, p. 542). The 'city' is not one homogenous, identical space. Cities differ from each other and within their own (permeable and unstable) boundaries they fragment and multiply—some becoming completely alien to others: conflicting and disparate. Cities are made up of contradictions and disjunctions, but the details of fragmentations are specific to particular locales. As Wasson notes, 'Increasingly, critics call not only for historicized but *localized* Gothic' (2010, pp. 3–4). *The New Urban Gothic* pays attention to these specific and particular locales, and its essays demonstrate the strong local characteristics of different and multiple urban Gothics that all reflect and report on the human residue and wastage of anti-landscape in the Anthropocene.

The book is divided into three parts: 'urban Gothic: Identities and Histories', 'urban Gothic: Ruin and Residue in the Anthropocene' and 'Global Gothic: Decentring the urban Gothic'. Each part looks at a different aspect of the new urban Gothic. Part One looks at vulnerable, rejected and, often, intersectional identities and their connection to the urban environment. The chapters in this part all recount how, residual, and 'peculiar' identities have a specific, Gothicised relationship to the spaces in which they are found and contained. Those who reside in the periphery of society also reside in the periphery of the city—a space they come to 'own' or stamp as territorially marginal; a melding of the body-of-the-city and the marginalised bodies on the periphery. Thus, we begin with these identities and their histories.

The first part begins with a retrospective on the urban Gothic by Julian Wolfreys. Wolfreys has graciously accounted for the phenomena that in the last few decades Gothic Studies have witnessed an increasing interest in revisiting, reproducing or rewriting various aspects of nineteenth-century culture and readers will recognise this trend throughout this collection. This representation of the urban has had particular traction in Gothic texts, especially since the postmodernism of the 1980s and early 1990s and the rise of Globalisation during this time. Karl Bell opens the essays in this part with '"Exquisite Corpse": The urban Gothic Mindscape in China Miéville's *The Last Days of New Paris*'. Bell's discussion juxtaposes Miéville's work with the Victorian poem 'The City of Dreadful Night' by James Thomson (1880). Exploring the psychogeographic horror and the claustrophobia of cities, Bell argues that the neo-Victorian text moves away from the nineteenth-century depiction of the 'hidden horrors' of the city to a space where 'horrors are made deliberately visible'. Citing the contemporary city as a space of terror(ism), Bell suggests that the new urban Gothic speaks to a very immanent sense of threat, battle and anxiety.

Michael Fuchs follows with his chapter, '"Things are not what they seem": Colonialism, Capitalism and Neo-Victorian London in *The Order: 1886*'. Fuchs examines how the British Colonial project is depicted in the videogame *The Order: 1886* (2015). A meta-text composed of various myths, tropes and Victorian legends, the game populates London with the undead and uses them to 'symbolize the vampiric nature of the colonial enterprise'. Presenting a narrative that critiques colonialism, Fuchs argues that *The Order* interconnects with current concerns about globalisation, its addicts and its residue. Ultimately the game shows the city to be an uncanny space which doubles the nineteenth-century setting with colonial practices and processes that are still in place.

In chapter three, R.M. Francis moves the urban location to the Black Country regions of Britain in his essay 'A Very Queer Black Country'. Reading the weird fiction of Joel Lane, this chapter looks at the portrayal of The Black Country and its characters, as examples of a post-industrial Gothic in flux, contradiction and transition. Arguing that the Black Country is a queer space which is borderless, both green and grey, and inherently contradictory, Francis looks at the alienated figure of the queer subject and the unearthing of liminal sexuality which this landscape makes possible. Francis sees the inherent otherness of the queer subject as finding some sense of recognition and acceptance in the decaying Gothic spaces of a Black Country on the edge of the millennium. Here, the sense of place, the borderlessness and geographical contradictions act, both as anti-landscape and as a means through which very particular genders and sexualities are forged or played with.

In the last chapter of this part, Kwasu Tembo writes of 'Abjection and Anime in the Anthropocene: Amano and Oshii's *Angel's Egg* (1985)'. Focusing on contemporary anime, Tembo looks at the aesthetic slant the film takes on the Gothic. Tembo suggests that *Angel's Egg* brings a surrealist aesthetic to anime that is unique. Locating his discussion within the context of a World's End tale, Tembo looks at how the aesthetic and characters portray different images of a post-apocalyptic, Gothic space that draws on, what he terms as, 'psycho-emotional Otherness'.

The second part—'urban Gothic Ruin' looks at urban space as destabilised and de-centred. In these chapters, the city is rendered uncanny either through a feeling of dislocation produced by a vertiginous shift between the real and the unreal (and then back), or by a sense of the dual nature of urban space in the doubling of above-ground and below-ground worlds. Concentrating on texts in different mediums—video games and novels, these essays look at the underside of the city and its material otherness. The part begins with an introductory essay by Holly-Gale Millette on the diachronicity of the dystopian aspects of our urban Gothic narratives in the Anthropocene. The specific focus of this part is on the dialectical relationship between the two (geologic and philosophic) historic periods, ruin and residue, and the conceptual traffic of the

Anthropocene as accounted for in millennial urban Gothic texts. Importantly, it takes for granted that which all who argue and debate the subject of the Anthropocene can agree with: (1) that we are consciously aware that vast geologic events and changes—such as the degradation of mass ecosystems on a planetary scale—are absolutely connected to what the human species does in their everyday lives, in their multitudes and collectivities, their institutions and their global nation-states; (2) that these acts and systems are dark and monstrous in the Gothic sense and (3) that, systemically, the impact of the Anthropocene is most visible in the urban but this is no longer confined to the canonically urban locations of New York, Paris and London.

The first chapter in this part is Garth Sabo's '"A Weapon in the Cracks": Wasteways Between Worlds'. Looking at the underside of the city, Sabo explores the material filth, refuse and decaying matter that every city produces. Analysing China Miéville's *The City and the City* (2009) and David Mitchell's *Slade House* (2015), he traces 'the dynamics of street rubbish', arguing that both narratives suggest that we ignore this residue at our peril. Placing the texts in a larger context of impending ecological disaster, Sabo examines the Gothic 'irruptive force of discarded matter' and 'the urgent materiality of waste' in contemporary cityscapes. As he says; 'Trash destabilizes these novels, making it difficult, if not impossible, to imagine ontological spheres that can ignore or avoid contemporary ecological crises'.

Next, Madelon Hoedt turns to video games in her chapter: '"Lord, What a Splendid World We Ruined": The City and the Underground in *Metro 2033* and *Metro: Last Light*'. Hoedt looks at the setting of a 'new' Mosco in these videogames and the hostile environment depicted both above and below ground. Presenting a double-city, players navigate a traditional space that has been inverted such that the underground space becomes more familiar than the ruins of the above-ground city. Drawing on theory from Gothic scholars about the 'fear of what lies beneath' (Mighall 2007), Hoedt looks at the fascinating way these popular videogames have turned this concept upside down in the wastes of a world turned upside down.

Staying with games, Tanya Krzywinska's chapter '"Everything is True" A Weird Tale: urban Gothic meets Urban Myth in the Multiplayer Online Game: *The Secret World*', argues that a new aesthetic has emerged within multiplayer online games enabled by new technological capabilities that produce an 'edgy and contemporary urban Gothic'. These games relocate the action to the here and now—to a 'realistically rendered world that calls on our own time and space'. Krzywinska focuses on the horror game *The Secret World* (2015–present) which has settings based on real-life cities. This creates a 'strong sense of the vertiginous nature of the Weird' as there is what she calls 'a psychoactive blend of urban Gothic and Myth'. Krzywinska suggests that this melding of the real with the gaming persona and the very unreal nature of the actual game cultivates a sense of unease and uncertainty that has moved gaming on to something new.

The ruinous post-industrial landscapes of Detroit are the setting for Leila Taylor's chapter: 'The Rust Belt of Ruin and the Gothic Genius Loci of Detroit'. Citing the contemporary cityscape of Detroit as relative to what she terms the 'ruin porn' of Gothic topography, Taylor focuses on two films *Only Lovers Left Alive* (2013) and *It Follows* (2014). She argues that Detroit is now valued for its decay and that there is a Gothic pleasure to be had in the representation of the decline and failure of capitalism that this urban landscape materially signifies. Like the ruined Abbeys or castles valued by Gothic tourism, Detroit's urban wilderness shifts 'into the realm of aesthetics and romanticism'.

The final part of the book 'Global Gothic: Decentring the urban Gothic' is introduced by Ruth Heholt. Beginning with an examination of a new transnational turn in Victorian studies, this piece traces a globalisation of the Gothic that seems equivalent. Heholt argues that the Gothic has always been global and specific to place as well as time. While in many discussions of the Gothic there is a tendency to concentrate on the British (first), the European and then American texts, forms and tropes, other parts of the globe have been nurturing their own Gothics. These non-Eurocentric expressions of the new urban Gothic encompass multiple points of view. The prolific, fertile and plural Gothics from around the world are explored in the chapters in this part. The discussion begins with Katarzyna Ancuta's chapter, 'Communal After-Living: Asian Ghosts and the City'. This chapter traces the ghosts that are left behind in Asia's communal high-rise buildings, identifying what Ancuta calls 'Asian apartment horror'. These apartments are spaces where the living and the dead exist side by side. Characterised by loneliness and isolation, in these places the 'ghosts are more visible than the living'. Looking at films from several different countries, Ancuta reads them and their ghosts as representations of a 'failed dream of economic success' that drove people into the cities from the rural districts in the first place. Offering a uniquely dystopian take on the haunted house movie, the films examined in this chapter express a deeper sense of failure, displacement and invisibility in spaces only found in cities.

Molly Slavin takes us to India in her chapter: '"In Those Days the Little City of Srinagar Died with the Light": *The Ministry of Utmost Happiness* as New urban Gothic'. Slavin looks at the way in which Arundhati Roy demarcates the cities into areas—the Colonial and the Postcolonial. Arguing that she deliberately writes her protagonists as 'out of place', Slavin follows Roy's protagonists as they 'reinvent Gothic tropes for the twenty-first century'. Questioning ideas of home, Slavin argues that the novel is an example of the urban Gothic trope that enables the exploration of contemporary Indian urban issues such as inequality and violence. With an unusually optimistic tone, Slavin argues that '*The Ministry of Utmost Happiness* uses the new urban Gothic sites of Delhi and Srinagar to translate Colonial and Gothic conventions for productive, justice-oriented purposes in contemporary South Asian cities'.

Taking something of an autobiographical stance, Gina Wisker's chapter, considers Singapore as a Gothic urban space. Her chapter, 'urban Gothic: Singapore' examines the underside of the glittering, capitalist, competitive face of the city. Arguing that there is a hidden, ghostly, resentful 'parallel world to the that of the managed, outward facing' side of the city, Wisker suggests that it is the darker Singapore that has more to tell us about urban life. Looking at work by Sandi Tan and the film *Crazy Rich Asians* (2018), Wisker explores the history and the underside of a Singapore which is rarely mentioned or acknowledged in public. These texts speak of 'the haunting presences and howling silences of ghosts of the bustling past whose dark secrets have been nearly erased, but [which still] exist parallel to the present, imprinted on the spaces, places, buildings and in the living histories of those who inhabit the island city'.

Anne-Marie Lopez's chapter, '*Suzhou River*: On the [haunted] Waterfront' continues the examination of rapid economic growth and the 'social and aesthetic malaise' wrought on the urban environment and the people who inhabit these spaces. Looking at *Suzhou River*, Lopez traces the narrative and aesthetic legacy of a film that was banned from screenings in China. Crossing genre boundaries, *Suzhou River* leaves the viewer unsettled as the narrative viewpoint shifts and what is presumed to be a stable form—the documentary-style of an over-arching narrative voice and the first-hand tale of a videographer—is destabilised. Lopez shows how through a 'disjunction between visual evidence and fictional constructs', the film weaves the viewer 'into its spectral world of ghosts and doubles'. Blending Eastern and Western conventions of horror and the Gothic; engaging with Hitchcock's film *Vertigo*, *Suzhou River* presents a 'transformative imitation' that echoes and reflects 'Chinese hybridity', while also commenting on the failure of 'propaganda about Shanghai's bright future'.

In the final essay of the book, Xavier Aldana Reyes takes a timely look at Barcelona Gothic. Examining Carlos Ruiz Zafón's *The Cemetery of Books* (2001–2016) series, Reyes shows how a retrospective look back at the Spanish Civil War provides a Gothic corrective to the tourist vision of the city. Looking also at *Marina* (1999) and the collection *Barcelona Gothic* (2008), Reyes suggests that there is a pull between a desire to create a regional Gothic specific to Barcelona and a desire to 'explore the legacies of the Spanish Civil War'. Reyes argues, finally, that Zafón manages to blend the two producing a Barcelona Gothic as metatext, that exists *because* of a past that both haunts and marketises the city.

All the chapters look at specific places of post-industrial ruin and their relationship to the past, to trauma and the emotional pull of spaces of decay. Sometimes seeing these anti-landscapes as providing space for those who can never 'fit' into the light, bright city, the essays in this book look at the possibilities presented by a dark, despoiled city. Here an acceptance of human trauma, violence and destruction produce Gothic spaces. Contextualising the past in a direct and immanent relation to present inequalities and social, urban problems, the essays in this book bring the new urban Gothic right up to the present moment. Overall, writing in *The New Urban Gothic* links the idea of the Anthropocene to older Gothic tropes as well as those found in the new and emerging urban storyworlds. Examining the different and shifting meanings that arise from contemporary fictions that use the legacy of Victorian urban Gothic in their storytelling, the volume looks at the intersections of time, place, space and media in a Global context. It's a distinctive collection for how it casts reflections and shadows on the age of the human and the deconstruction of ideas of progress, evolution and enlightenment.

The New Urban Gothic is invested in human-centred temporalities, not planet-centred ones. This collection recognises that the history of the human species and the history of the planet do not behave similarly in context or situation (Zalasiewicz 2017), and yet geologic time and philosophic human time are certainly reliant on one another as geologic time is organised from the prima-facia Cartesian understanding that 'man (and his/her presence on the planet) is the measure of all things'. These understandings are neither simple nor unsituated and, increasingly, critical post-humanism has us mocking the Cartesian ideal as hubristic. This is not a volume dedicated to World's End expectations, although it is true that World's End narratives and critique do abound. Instead, this book focuses on the tension that exists between dark experiences and darker future expectations. Indeed, the very tension between experience and expectation gives rise to patterns of behaviour and prediction that mean historical time cannot be separated from human affect. As Christophe Bouton has pointed out in his commentary on Koselleck's categories of our capacity to die, our capacity to kill and the threat of violent death as the background of any history: '[p]rospects of the future, raising hopes and anxieties, making one precautionary or planful' are basic to the transcendental structure of history and without these—without the capacity to kill one another—'the history we all know would not exist' (Bouton 2016, p. 178). At its most base, this is what is at stake in the Gothic of the Anthropocene. This and the violence that it evolves from.

'World's End' narratives that proliferate on screen and in our culture are the most violent of these Gothic tales. We are culturally imbued with the knowledge that '[m]ankind is becoming a force in climate comparable to the orbital variations that drive glacial cycles' (Archer 2009, p. 64) and that 'humanity is drifting into unparalleled catastrophe' (Glikson and Groves 2016, p. 192) of the Species End variety, a process that Glikson and Groves call 'Planeticide'. The prospect of such a planetary or species level extinction is the darkest of Gothic tales, precisely because it has never happened before and would be the most monstrous of cataclysms. 'If human activities eventually lead to a Sixth Great Extinction of the species, it [will] be a "first" for the planet' cautions Chakrabarty (2018), 'Never before has a species caused a mass extinction event' (p. 26, ftn. 86). Now *that* is dark; the darkest of times. Previous extinctions were all caused by 'some strong combination of asteroid impacts, clusters of volcanic eruptions, ice ages, and/or indications of large changes in ocean chemistry'. (Reznick 2010, p. 310). And yet here we are, staring down the barrel of mass extinction after a mere few hundred years on the planet—a fraction of the time the dinosaurs spent inhabiting it before being destroyed by a meteor. Only the human monster has succeeded in living so fast and so cheap that we are on the precipice of extinguishing our species: Well done us.

Note

1. It is a basic Benjaminian tenet that the affairs of the living are aesthetic—structured as an art form—and that, conversely, because everything in life is political, politics are likewise structured as an art form. This entwined system he discusses best/first in *The Work of Art in the Age of Mechanical Reproduction*.

References

Agamben, Giorgio. 1993. *The Coming Community*, trans. Michael Hardt. Minneapolis, MN: Univeristy of Minnesota Press.

Archer, David. 2009. *The Long Thaw: How Humans Are Changing the Next 100,000 Years of Earth's Climate*. Princeton: Princeton University Press.

Bauman, Zygmunt. 2012. *Culture in a Liquid Modern World*. Cambridge: Polity Press.

Beck, Ulrich, Anthony Giddens, and Scott Lash. 1994. *Reflexive Modernization: Politics Tradition and Aesthetics in the Modern Social Order*. Cambridge: Polity Press.

Benjamin, Walter. 1969. The Work of Art in the Age of Mechanical Reproduction. In *Illuminations*, trans. Harry Zohn. New York: Schocken Books.

———. 2003. *Walter Benjamin: Selected Writings*, ed. Marcus Bullock and Michael Jennings. Cambridge, MA: Harvard University Press.

Bouton, Christophe. 2016. The Critical Theory of History: Rethinking the Philosophy of History, in the Light of Koselleck's Work. In *History and Theory* 55: 2.

Chakrabarty, Dipesh. 2018. The Seventh *History and Theory* Lecture: Anthropocene Time. In *History and Theory* 57, no. 1 (March): 5–32.

Clover, Carol. 1992. *Men, Women and Chainsaws: Gender in the Modern Horror Film*. Princeton: Princeton University Press.

Crang, Mike. 1998. *Cultral Geography*, ed. David Bell and Stephen Wynn Williams. The Routledge Companion to Human Geography. Oxon: Routledge.

Glikson, Andrew, and Colin Groves. 2016. *Climate, Fire and Human Evolution: The Deep Time Dimensions of the Anthropocene*. Cham: Springer.

Harvey, David. 1996. *Justice, Nature and the Geography of Difference*. Oxford: Blackwell.

Horsley, Lee. 2005. *Twentieth-Century Crime Fiction*. Oxford: Oxford University Press.

Jackson, J.B. 1984. *Discovering the Vernacular Landscape*. New Haven: Yale University Press.

Jameson, Fredric. 2013. *Antinomies of Realism*. London and New York: Verso.

Jenkins, Henry. 2009. Transmedia Storytelling. In *Volume* 19: 56–59.
Lefebvre, Henri. 1991/1974. *The Production of Space*, trans. Donald Nicholson-Smith. Oxford: Blackwell.
Luckhurst, Roger. 2002. The Contemporary London Gothic and the Limits of the "Spectral Turn". *Textual Practice* 16 (3): 527–546.
Massey, Doreen. 1994. *Space, Place and Gender*. Cambridge: Polity Press.
Mighall, Robert. 2007. Gothic Cities. In *The Routledge Companion to Gothic*, ed. Catherine Spooner and Emma McEvoy, 54–62. Abingdon: Routledge.
Moore, Jason. 2015. *Capitalism and the Web of Life: Ecology and the Accumulation of Capital*. London: Verso.
Nancy, Jean-Luc. 2000. *Being Singular Plural*. trans. Robert D. Richardson and Anne E. O'Byrne. Stanford, CA: Stanford University Press.
Nixon, Ron. 2011. *Slow Violence and the Environmentalism of the Poor*. Cambridge, MA: Harvard University Press.
Nye, David E., and Sarah Elkind. 2014. *The Anti-Landscape*. Amsterdam: Rodopi.
Rebellion.earth. Accessed 3 December 2019.
Reznick, David N. 2010. *The Origin: Then and Now: An Interpretive Guide to the Origin of the Species*. Princeton: Princeton University Press.
Warwick, Alexandra. 1999. Lost Cities: London's Apocalypse. In *Spectral Readings: Towards a Gothic Geography*, ed. Glennis Byron and David Punter, 73–87. Houndsmills, Basingstoke: Macmillan Press.
Wasson, Sara. 2010. *Urban Gothic of the Second World War: Dark London*. Basingstoke: Palgrave Macmillan.
World Health Organisation. https://www.who.int/airpollution/ambient/en/. Accessed 3 December 2019.
Zalasiewicz, Jan. 2017. The Extraordinary Strata of the Anthropocen. In *Environmental Humanities: Voices from the Athropocene*, ed. S. Oppermann and S. Iovino. London: Rowman and Littlefield International.

Urban Gothic: Identities and Histories

Julian Wolfreys

At some point, sooner or later, the idea of the city, and the many and varied modes of its representations, would be presented in terms of Gothic. These have areas of similarity, areas of overlap. They also have areas of difference from one another. As such, Urban Gothic (and by extension and form of Gothic or 'new'-Gothic never entirely like itself, but also never entirely different), will strike a chord with viewer, reader, game-participant or audience alike. Gothic—no definite article, for to use this would be to delimit implicitly the multiplicity, the 'swarms' of Gothic manifestation and mutation and therefore domesticate and familiarise what is already in some cases overdetermined in controllable ways—is *not one*: always in the process of 'othering' itself, Gothic transforms and is transformative. There is no straightforward 'genetic' code for Gothic that is not, always already, corrupted, auto-evolving, within whatever might momentarily be perceived as an 'itself', a stable and recognisable concept or ontology, with an equally stable epistemological framework and embedding. Customary or conventional as it is in criticism to employ a capital G in referring to the cultural expressions of Gothic—film-text, literary text, game-text, painting, graphic novel, etc—ironically, and perhaps not a little paradoxically, this very convention capitalises on nothing so much as the fact that Gothic is irreducible to one form, one identity. This *is* Gothic's identity: to have no one form, at the risk of repeating myself.

This is not to say anything remarkably new or even unknown. Today, many critics and commentators on all forms of Gothic take this as read, as given, in order to get on with business as usual. However, in order to begin again as it were, in order to locate an initial departure point in the liminal and crepuscular psychogeography of Gothic in general and especially in the landscape and cityscape of the Urban Gothic and new Urban Gothic, it is good to remind ourselves of just how wayward Gothic is, and how such auto-transgressive resistance to absolute familiarisation can inform and so promote and encourage the representation of alternative identities and so give articulation to the many articulations and affirmations of the Gothic subject, queer, postcolonial, LGBT or any other 'other'.

This is though not yet to speak of the Urban, at least not explicitly. Again, like the idea of Gothic, there is no one Urban. If anything, we accept this much more readily, if not uncritically, than we do with Gothic. Without offering so many 'variations on a theme', allow me to suggest that we take it as given that Urban is, like Gothic, a complex and heterodox pseudo-portmanteau term, with little to its understanding that could be categorised as a normative taxonomy: like Gothic, heterogeneous within the 'itself' that the term implicitly signals, the many 'selves' of Urban, sometimes touching on, informing and coming to be informed by the many 'selves' of Gothic (and so producing or inventing ever more fascinating, seductive, dangerous and to some, threatening creatures), resists complete normalisation and complete comprehension.

Accepting this, I would argue that it is the impossibility of a full comprehension that lies at the root of the reading subject's apprehension in the face of Urban Gothic in general, and New Urban Gothic in particular. Such apprehension is realised in that the texts of new Urban Gothic all produce, each after their own fashion, subjects-in-apprehension, each embodying, enacting, performing, articulating and affirming their Being-in-apprehension. Being dwells or might only have the possibility of dwelling authentically in new Urban Gothic, to be explicitly Heideggerian, on the condition that it realises in its actions, perceptions and reflections on selfhood and the self's place in urban landscapes—and as a reflective and meditative condition of those constructed or built sites—to the fullest extent how the 'nature' of Being-as-dwelling is fundamentally to be always already homeless, ***unheimlich***, uncanny. Understanding this as an explicit function of the new Urban Gothic is important. Its importance resides in a fact that is as materially and culturally historiographical as it is phenomenological and ontological.

To explain this as directly impossible: if one imagines, in an act of retrospective construction, a narrative trajectory for Gothic and Urban

Gothic, then taking as a well-acknowledged inaugural moment—fiction or myth of origin—the conventionally agreed a 'first' Gothic narrative such as Horace Walpole's *The Castle of Otranto* (1764) and subsequent novels or romances from the first flush of Gothic such as Ann Radcliffe's *The Castles of Athlin and Dunbayne* (1789), and even Jane Austen's parody of Gothic *Northanger Abbey* (1817),[1] what is blindingly obvious about such texts—what they have in common—is *place*. Each novel names as its principal setting a built environment. So far, so obvious. And each structure determines in large part the eventual determination of the novel's principal subject. Place and self, Being and dwelling, Being and the uncanny: all are inextricably realised in literature as mutually interdependent in a way that arguably had not been foregrounded before. Moreover, this is not a simple one-way street. Setting does not determine subjectivity alone. The subject, fully engaged in the culturally overdetermined anticipation and phenomenological perception of a place, gives to a specific location through reading of architecture, ambience, tone or mood (perhaps best described by the German term, *Stimmung*) a mode of presentation, a staged 'identity'. Place thus becomes an entity in its own right, imbued not only by the author but in the perception of the subject-reader and the reader of the text (the two subjects are often aligned to a degree to the extent that the author withholds or discloses information or otherwise determines reception through the rhetorical and poetic modes of perceptual production, in order to make material the 'affect' of place) with an uncanny sense of proximal Being.

Thus, there came about—to risk a hypothesis—early explicit presentations of both psychogeography and psycho-architecture, both in turn, as determinants of the content, serving to shape the *psycho-archetectonics* of Gothic narrative, Gothic poetics and their subjects. The structure and trajectory of the narrative was thus constituted through the interaction of Being and Dwelling, struggling towards revelation and reconciliation, in which the psychology of self and its many archaic and irrational fetishes came to be foregrounded through narratives of mystery and solution of those mysteries. Or to put this differently, Gothic narrative and poetics involve in its first acknowledged forms both a psychology and a psychoanalysis of subjectivity; and this subjectivity was in this narrative unfolding understood by the readers of Gothic as *modern*, like themselves—or at least, as they would have liked to have imagined themselves, eventually rid of any atavistic, fetishistic or superstitious residue belonging to earlier historico-cultural manifestations of the self. Moreover, such a gradual revelation of modern subjectivity, which necessarily took time inasmuch as it had never before been so directly articulated and reflected upon to such a great and sustained extent, required two elements: on the one hand, the temporality of an exploratory narrative that, like many adventure and puzzle-solving computer games or the past couple of decades, took time to develop and evolve in order that the subject might not only banish all that was initially accepted as frightening or unfamiliar, unknown, foreign, other, but also would come to arrive at a greater understanding of him- or herself. On the other hand, Gothic narrative needed the materiality of place not only as a setting, but specifically as the ground of the self, on which the subject could be constructed and so be given life—not unlike Victor Frankenstein's creature, who, today, as I write, has returned via Jeanette Winterson's latest novel, *Frankisstein* [2019] concerned with LGBT and transgender identities, the question of AI and the scientific possibilities of gender reassignment, all of which is given, in part, the setting of an increasingly Gothic Brexit landscape, where Englishness is, itself, an uncanny and monstrous manifestation.

Of course, Winterson does not need to provide an extended temporality of selfhood, as writers of what we might conveniently term 'first wave Gothic' had to, if only so as to begin to understand themselves the ways in which place and Being were closely entwined. She can foreground different modes of Being, other stagings of identity and selfhood differently Gothicised and urbanised, as can writers such as Peter Carey or Robin Robertson, in texts as wildly different as *Jack Maggs*, the postcolonial re-envisioning of *Great Expectations*, with its dark and forbidding, crepuscular London, or *The Long Take*, Robertson's bleak take on the post-war deconstruction of the self in a mid-century staging of Los Angeles that is as Gothic as it is hard boiled, owing as much to the films of Billy Wilder and Fritz Lang as it does to the fictions of Raymond Chandler or James M. Cain. Or, to put this differently, Winterson or those authors, filmmakers and game designers considered in the present volume each work in temporality, in temporalities, that are markedly different from those of first, second or third generation Gothic writers and artists of the Urban Gothic, from Walpole to Dickens, from Brandon to Stevenson, from Wilde to Elizabeth Bowen's short stories, and beyond. And here is the argument I have begun to develop. Over the history of Urban Gothic textuality, following the 'first' efforts to bring to consciousness the relation between self and place as a modern phenomenon, subsequent narratives (from Austen onwards) have come to terms more and more rapidly with this intimacy and mutual interdependence, thereby making telegraphic the work needing to be done, while also domesticating Gothic in moving event and staging to seemingly familiar locations of the subject's own culture, nation, home. Furthermore, due to the attenuation of the once necessary temporal space of earlier Gothic narratives, present-day filmmakers, game designers and writers can explore, with all due acknowledgement of and reference to the poetics, rhetoric and politics of Urban Gothic, as this has been codified and made phenomenally and materially present over a roughly 250 year period, the transformations of urban space in post-industrial, postcolonial and post-heteronormative contexts, in rich and surprising ways. As the authors of the diverse essays in this section demonstrate, the latest generations of writers, artists, filmmakers, game writers have continued to expand and explore different cultures and sites of Gothic, and to reimagine those sites in inventive ways, even to the extent that there is already underway, in a global reimagining of the Urban (new) Gothic, a revision of political and poetic purpose in what just over a generation was still somewhat new, but now appears as a canonical genre, the neo-Victorian.

What this collection of essays makes abundantly clear is that there is no shortage of imagination when it comes to either the urban or the Gothic. We dwell, as unhomely as the experience may be, in both, the modernity of our Being determined by a dwelling that is also a displacement, homelessness; and with this arrives, as the authors gathered here demonstrate, a persistent sense of anxiety and the uncanny, as each and every modern urban subject remains caught up in, and in the wake of a search for identity, subjectivity and selfhood. We might even say that what the modern new Urban Gothic can teach us about ourselves is that, always already estranged from any home, we each of us travel in the wake of the ghosts of ourselves, those phantoms and spectres we believe ourselves to be. Never catching up with such haunting figures of alterity that lies at the heart of who we are, we, like those subjects of the earliest Gothic texts remain in thrall to what is closest to us: our others, our selves. The landscapes we and they inhabit are those of our own waking dreams.

Note

1. I would like to propose briefly a hypothesis, which must remain so as I do not have the space to explore in rigorous detail must remain for now merely a suggestion. While Austen's novel is parody and satire, a comedy

of manners and a spoof on what was by 1817, already becoming an all too familiar and domesticated genre (Gothic); and while, a novel produced a year later, Mary Shelley's *Frankenstein*, was received as being shockingly new and, as its subtitle suggests, 'modern' (both modern and ancient, in that this is the tale of the 'modern Prometheus'); and furthermore, while Shelley's text has in the past 50 years or so been endlessly reinvented as germane to our various and different historical present or read critically for the ways in which it can be taken *avant la lettre* as recuperable so as to make comment on whichever present the critic might inhabit, It is, I would like to argue (perhaps outrageously as some might see it) that Austen's novel is the more 'modern' of the two. For it reads subjectivity as both phenomenological (Catherine Moreland perceives the world through the filter of literary mediations and so misperceives both the world and herself, only gradually coming to consciousness of her selfhood and identity) and materially determined—Catherine only has self to the extent that she is materially and culturally embedded in the world (she is an early expression of Being-in-the-world, of Being-there, as opposed simply to merely Being in some non-materialist, ahistorical post-Platonic or Post-Cartesian manner). While what Shelley has to say about the ethics of science, which remains relevant today, selfhood in Austen's understanding is articulated both philosophically and materially, and is thus a necessary 'modern' precursor to manifestations of both psychoanalytic and phenomenological thought as these figure perceptions and constructions of the self-in-the-world, as this is constituted through the interaction of self and place and reading (or initially misreading, which is crucial) by self of place.

Exquisite Corpse: The Urban Gothic Mindscape in China Miéville's *The Last Days of New Paris*

Karl Bell

During the Paris riots of May 1968 the city's walls were adorned with slogans declaring 'Sous les paves, la plage!' ('Beneath the paving stones, the beach!'), and 'La poesie est dans la rue' ('Poetry is in the street'). Such statements appear to have been marled into China Miéville's world building in *The Last Days of New Paris* (2016), a novella that imagines the fantastical ecological re-invention of the Parisian cityscape. Situating James Thomson's hypnogogic poem, *The City of Dreadful Night* (serialised 1874; compiled 1880), as a Victorian urban Gothic predecessor, this chapter explores how Miéville's text contributes to the new Urban Gothic through its depiction of the ecopsychological externalisation of inner mindscapes. In doing so, it fuses Gothic criticism's familiar emphasis on psychological interiority with its more recent turn towards historicist approaches and environmental understandings.

This chapter begins by briefly outlining the longer Gothic tradition which linked emotions to external landscapes. It then teases out resonances between Thomson's narrative poem and Miéville's novella,

K. Bell (✉)
University of Portsmouth, Portsmouth, Hampshire, England, UK
e-mail: karl.bell@port.ac.uk

H.-G. Millette and R. Heholt (eds.),
The New Urban Gothic, Palgrave Gothic,
https://doi.org/10.1007/978-3-030-43777-0_2

focussing on the way their cities are transformed by mood and imagination, and by spatial and temporal distortions that render their London and Parisian locales *unheimlich*. Having considered some of the ways Thomson's poem haunts Miéville's text, it then explores how New Paris both departs from and adds to the urban Gothic tradition. Its focus on Paris in the Second World War highlights the importance of considering the urban Gothic in other historical periods and places beyond its Victorian and Anglo-centric roots. It also examines connections between the Surrealist movement and the Gothic, most obviously in the importance Surrealist artists and Gothic scholars have granted to Freud's work on the uncanny. The chapter concludes by exploring how the Surrealists' revolutionary politics and aesthetics find expression in the urban grotesque of New Paris, and how the new urban Gothic might also articulate our contemporary political and environmental concerns in the age of the Anthropocene.

Thomson and Miéville's texts can be situated within a long Gothic tradition of external landscapes reflecting disturbed interiority, heightened mental turmoil and fearful imaginings seemingly made manifest in one's physical surroundings. This can be traced back to works that pre-date the emergence of Gothic literature, most obviously Edmund Burke's *A Philosophical Enquiry into the Origins of Our Ideas of the Sublime and the Beautiful* (1757). In his reflections on the sublime Burke established the relationship between perception and its physical and mental responses, with the vastness or might of nature inclined to generate feelings of terror and human fragility. Gothic writers of the late eighteenth century furthered this connection between the external environment and internal feeling, although they differed markedly in interpretation. While Anne Radcliffe's work presented nature as 'a conduit of emotions, a way to experience feelings and sometimes to purge them', a responsive environment that encouraged the mind to turn towards the divine, it was Matthew Lewis' more negative view that triumphed. He considered nature 'a feral and wild place', its isolation enabling humanity to act upon its most corrupt and depraved instincts without restriction (Kroger 2013, pp. 19–20). Developing this further, Ruth Heholt has noted how the boundary between inside and outside collapses in the Gothic landscape; '[T]hrough affect, intensity, heightened emotion, darkness and desire [it] conflates and collapses any differentiation between self and place' (Heholt 2016, p. 235).

Responding to the 'spatial turn', urban Gothic criticism has attempted to move away from a (predominant) emphasis on psychological interiority to appreciate the importance of spatialised and historicised understandings (See Phillips and Witchard 2012; Mighall 2003) in considering the way specific social and environmental factors play an important role in the creation of an Urban Gothic milieu. As will be indicated below, Urban Gothic texts that present the externalisation of psychological interiority can be seen to combine those differing approaches. Such texts confront us not with the horrors of repression but expression. The usually hidden and haunted becomes the very landscape through which the protagonists move. In such circumstance, the Gothic environment is no longer just a physical metaphor for the disturbed mind of the unreliable narrator; it becomes urban architecture, ecology and atmosphere.

Thomson and Miéville's works both read like hypnagogic experiences, a traversing of an urban dreamscape narrated by lone and, in different ways, lost individuals. Thomson's *The City of Dreadful Night* is a poetic account of a wandering insomniac in a perpetually nocturnal city. He encounters a range of solitary inhabitants, each locked within their own dramas or tragedies, worn by melancholy, wounded by loss, or desperate with despair. Each emerges from the gloom like some captive of an urban hell, and Thomson makes conscious allusions to Dante's *Inferno* to drive the point home (Thomson 1993, p. 40). Possessing the disjointed and cloying feel of a nightmare, the whole city is oppressed by a pervasive sense of sadness and pessimism. This culminates at the poem's end with the narrator coming before 'the bronze colossus of a winged woman' (p. 69); *Melencolia*, the urban goddess that embodies the spirit of Thomson's dreadful city.

Miéville's novella intercuts between two narratives in an alternate history in which the Nazi occupation of Paris continued to 1950 as a result of the catastrophic S-blast, an explosion that caused the materialisation of 'manifs', entities drawn from Surrealist art and writings. In one narrative, Thibaut, a member of the Main à plume, the Surrealist resistance, seeks to escape the city. Saving Sam, a photographer, from her Nazi pursuers, he is drawn into uncovering a secret Nazi project codenamed Fall Rot. The second narrative gradually reveals the origin of the S-bomb in 1941. Jack Parsons, an American disciple of the occultist Aleister Crowley, comes to Europe intent on animating a Golem to fight the Nazis. Finding himself in the company of leading Surrealists in Marseilles, he fills his occult machine with their games and images, drawing power 'from the hex-fouled land outside' (Miéville, p. 87). The machine is stolen, and when it is accidentally detonated in a café in Paris it recreates the city as New Paris.

While markedly different in tone and presentation, these texts nevertheless share certain resonances. Both Thomson's unnamed city and New Paris are twisted doppelgangers of London and Paris, transformed by significant spatial distortions that defamiliarise them from their respective nineteenth and twentieth-century realities. Thomson's city is located near a classically Gothic terrain of 'savage woods, enormous mountains … black ravines' (Thomson, p. 30), but his poem clearly evokes the mental anguish of the urban experience in Victorian London, albeit one that removes the familiar landmarks by which to recognise it. In enacting a loss and replacement of architectural signifiers, it leaves only the mental impressions of life in the later nineteenth-century metropolis. Miéville's New Paris also warps and re-estranges the city's known features through its Surrealist alterations, creating, as will be seen below, an entirely new urban ecology. This estranging is best symbolised by New Paris' Eiffel Tower, the top half still hanging suspended in the air, the bottom half having vanished (Miéville, p. 11).

These spatial distortions are accompanied by temporal distortions too. Thomson informs us that 'the sun has never visited that city' and that it dwells in an unending 'mausolean night' (Thomson, pp. 29 and 48). In a poem about the loss or abandoning of hope, Thomson does not grant his narrator any release through the easy symbolism of dawn. New Paris is distorted by its obvious divergences from known history. Its two chronological narratives meet in an anomalous temporal–spatial node; the café where Parson's occult device was detonated continually explodes, reforms and explodes again in unending succession. Caught in a loop, the café at the heart of the blast echoes the abolition of temporal progress seen in Thomson's poem.

New Paris is haunted by the wartime history of Paris that it overwrites. This haunting is enhanced by Miéville's inclusion of historical figures within the narrative, including many of the leading Surrealists, rocket engineer and occultist Jack Parsons, and the notorious Nazi doctor, Josef Mengele. These dissonances disturb and disrupt the reader's sense of historical familiarity. In New Paris, both mimetic reality and its history become the ruin left by the eruption of the surrealist imagination. Dropping the reader straight into this alien history, leaving us to gradually acclimatise to the newness of New Paris, it is only in the afterword that Miéville dispels this affect by relocating the focus to our own timeline. Here he introduces a familiar Gothic device in which the novella is revealed to be his version of a jumbled narrative told to the author in a hotel. This affects an updating of the 'found' manuscript that originally framed the presentation of Horace Walpole's *The Castle of Otranto*. While this coda reveals that the events in the story were not the last days of New Paris, no explanation is given as to what happened to it. If the narrator in the hotel is the aged Thibaut then he must have been able to cross into the (fictionalised depiction of the) author's timeline, yet the history of New Paris after 1950 is left unknown. Like Thomson's city, New Paris is left suspended in time, like the 'Great ruins of an unremembered past' (Thomson, p. 30).

Despite these distortions, both cities still conform to a familiar urban Gothic trope, that of the incarcerating edifice. Although alluding to a landscape beyond, Thomson's city appears to have no point of egress. In the echoes between its opening and closing stanzas, Thomson's poem forms a circular structure in which the anguished narrator appears destined to trek in an endless circle of damnation from which daylight or walking will never release him. The poem, like the consciousness of the insomniac it depicts, is itself an inescapable architecture. New Paris provides a more explicit example of what Sara Wasson has termed the 'carceral city' (Wasson 2010, p. 52). Enclosed by a physical barricade, guarded by armed soldiers, the Nazis keep the rival factions of resistance fighters, Free French and ordinary Parisians, stranded SS and German army units, manifs, and Nazi-summoned demons imprisoned within the transformed city. Thibaut reflects that 'The Nazis will never allow Paris to contaminate France. All roads in and out are locked down' (Miéville, p. 15). Even passage between the city's arrondissements is marked by a change in air pressure, as if each is hermetically sealed (possibly in both meanings of the word).

Yet it is not so much the cities' architectural structures and physical obstacles as the narrator or protagonist's mentalities and inclinations that maintain their self-imprisonment. Thomson suggests the city infects its inhabitants. Its 'atmosphere is dark and dense', its dwellers

> Each adding poison to the poisoned air;
> Infections of unutterable sadness,
> Infections of incalculable madness,
> Infections of incurable despair. (Thomson, p. 58)

This contagion seems to infuse the narrator's perception of his urban surroundings, for 'sombre mansions loom immense and dismal, the lanes are black as subterranean lairs' (p. 34). The city's architecture both reflects and reinforces the defeated, melancholic mood of those trapped within its confines. As Jamieson Ridenhour notes, 'The residents of the City of Dreadful Night are of a kind with their metropolis: they are dark and ruined, damaged by the city in which they wander and transformed into figures both pitiable and horrifying' (Ridenhour 2013, p. 106). Thompson's poetic suggestion of psychological pain and loss of hope spreading like a contagion is eerily prescient of Bernard Stiegler's notion of disaffectation and psychic and collective disindividuation (Stiegler 2012, pp. 80–102). Although defined by Stiegler as a consequence of hyper-industrialism and capitalism's exhaustion, Thomson locates such symptoms of despair at a considerably earlier stage of industrial development, one that arguably marked the historical emergence of the Anthropocene.

While the individual agency of Thomson's urban inhabitants is drained by low energy, mental turmoil and sapping despair, most of those trapped within New Paris find their agency overwhelmed by the eruption of manifested art and marauding demons. Transformed from the crowded, urban Gothic Paris of Eugene Sue's *The Mysteries of Paris* (1842–1843) or Gaston Leroux's *Phantom of the Opera* (1910), New Paris is more sparsely populated, although we are assured that its streets 'throng' (p. 180) with strange, animated marvels.[1] Like Thompson's narrator, Thibaut also nurses mental wounds. Plagued by guilt after not joining his surrealist mentors in a daring attack that resulted in their massacre, he is further burdened by the loss of the rest of his resistance cell when they venture into the Bois de Boulogne to find out what happened to them. He struggles with the idea of being a deserter, of the difference he could have made if he had been with his comrades, yet 'When he tries to think of leaving, Thibaut's head gets foggy'. (p. 93) Ultimately he fails to leave New Paris, although his decision to do so is a more empowered and empowering one than the hopelessness that finally defeats Thomson's narrator.

Although unacknowledged in the endnotes in Miéville's novella, traces of The *City of Dreadful Night* haunt New Paris's urban landscape. Thomson's poem prefigures Miéville's conceits in several ways. Heralding Surrealism and the occult inhabitants of New Paris, Thomson's poetry portrays the physical manifestation of Gothic metaphor. Notable examples include 'A woman with a red lamp in her hand' that, as she draws near, is revealed to be 'her own burning heart' (pp. 37 and 38), and a huge wagon, 'a Fate-appointed hearse' that rumbles out of the night 'bearing away [...] the joy, the peace, the life-hope, the abortions of all things good which [...] have been strangled by that City's curse' (p. 46). In the poem's final section, at the fringe of the city, he presents the reader with the actualization (although not the animation) of Albrecht Durer's depiction of *Melencolia*, 'That solemn sketch the pure sad artist wrought [...] with phantasies of his peculiar thought' (p. 69).

Also prescient is Thomson's depiction of a surreal form of combat between iconic images. In section XX the narrator, lulled by sleep, is witness to an unseen conflict between the statue of a sphinx and a stone angel bearing a sword. Each time his eyes close there is a clash, and each time he awakes the stone angel is further broken until it finally lies shattered, its 'trunkless head between the monster's quiescent paws' (Thomson, p. 68). Due to the narrator slipping in and out of consciousness this combat takes place as a montage, another technique favoured by surrealist filmmakers. The defeat of the angel, a symbol of religious power, by a hybrid creature of classical mythology adds to the atheistic rejection of God championed by the poem's anti-sermon in sections XIV–XVI. This also carries into New Paris where the demons, iconographic representatives of hell, clash with the manifs, the nature and motives of which remain as unfathomable as that of the sphynx. As expressions of creativity, the product of psychoanalytical techniques that bypassed rational thought and tapped into the realm of the unconscious, the manifs represent a secular, solipsistic power; the elevating of human imagination to the status of the divine. In New Paris, the most powerful manifs tend to be the more famous Surrealist images, their champion being Max Ernst's elephant, Celebes. This suggests the manifs draw their power not just from the individual artistic imagination but their adoption into the collective imagination.

Thomson's narrator also presages Thibaut as an urban wanderer, both of whom tends to serve more as a witness to their remade cities than as active agents with a clear purpose. Alternating between urban observations and reflections, the narrator of *The City of Dreadful Night*'s only act of assertion is his narration, the telling of which seems to hold him apart from those who represent more woebegone examples of his own wretched mental condition. Suffering the loss of his Main à plume comrades, Thibaut also seems to drift, enacting a psychogeographic *dérive*. As he writes in his notebook, 'The mission is vacant' (p. 19). Drawn to sites of occult power, most obviously Les Deux Magots, the cafe where the S-blast originated, Thibaut and Sam guide the reader through de-familiarised urban spaces and experiences. Our understanding of both cities is skewed by the subjective perceptions of the narrators. Thomson's city is shaded by his narrator's pessimism and insomnia, while Thibaut's story is revealed as a second-hand account relayed to Miéville. As such, both texts relate acts of ambiguous urban mapping, their cities and their narratives possessing a fragmented and episodic dream logic that engenders uncertainty about what we are being told.

Despite these resonances between Thomson's Victorian poem and Miéville's postmodern novella, *The Last Days of New Paris* offers some variants on the Urban Gothic tradition and suggest ways in which it is continuing to evolve. The popularity of neo-Victorian re-imaginings has resulted in a predominance of nineteenth-century Urban Gothic histories. Yet as Sara Wasson has demonstrated, there are later periods of history which are equally rich in Urban Gothic potential, especially the home front experience in the Second World War (Wasson 2010).[2] While the Gothic can be understood as literature of excess (Botting 1996, p. 1), it could be argued that the Second World War exceeded even its most fearful imaginings. In scale and impact it possessed its own sublime affect, confronting people with a sense of their own insignificance, frailty and mortality. The wartime city saw the literalising of Gothic motifs: 'People were buried alive in their own homes, night streets turned into a bizarre dreamscape' (Wasson 2010, p. 4) and bombing de-familiarised urban surroundings into a ruined wilderness. At the same time, wartime behaviours such as blackouts, air raids and suspicion of spies created an atmosphere ripe with anxiety, dread and paranoia. Although distorted by its overtly fantastical aspects, Miéville's novella presents us with the urban Gothic of Nazi-occupied Paris.

Given the context of Miéville's New Paris, Mellor has also noted that wartime damage often created the strange and irrational juxtapositions commonly found in Surrealist art. He argues that 'a mode that had seemed outlandish and contrived now became an explanatory tool in unprecedented situations, enabling reportage when more conventional tropes appeared debased or inadequate …. The wartime city [became] filled with juxtapositions, resurgent vegetation, horrific tableau, and significant fragments' (Mellor 2011, p. 93). While the Gothic traditionally used ruins as a material signifier of a haunting past or 'the epitome of psychological, bodily ... and aesthetic breakdown' (Cavallaro 2002, p. 13), the focus on the urban destruction of the Second World War, especially its surrealist variant in New Paris, suggests they have an altered meaning in the new Urban Gothic. The S-blast both destroys and transform Paris, vivifying it into a surreal terrain of weird architecture and bizarre monsters, a warped and living ruin that combines and animates organic and inorganic matter in wilful resistance to logic or reason. In doing so, ruins are no longer signifiers of the past but indicators of how New Paris has fundamentally broken away from history; they become relics of a new present (in an alternate 1950) and the embodiment of a new, unknown (and never revealed) future.

Miéville's concentration of Surrealist images and imaginings draws attention to the movement's understudied Gothic grotesquery. Seeking to represent an otherworldly subconscious in an age of psychoanalysis and world wars, Surrealism's bizarre juxtapositions were intended to disrupt and de-contextualise, altering and estranging the common function and proportions of objects. The manifs form a grotesque urban ecology, becoming New Paris' flora and fauna. Thibaut observes that '[T]he grass underfoot is speckled with plants that did not exist until the blast: plants that make noise; plants that move', while in the mud of the river 'human hands crawl under spiral shells' and 'Seine sharks thrash up dirty froth.... In front of each dorsal fin, each shark is hollow-backed, with a canoe seat' (p. 11). Paris's famous landmarks are similarly transformed into surreal oddities, with the Sacré-Cœur becoming a living tram depot, the Arc de Triomphe a urinal and the towers of Notre-Dame industrial silos, one filled with blood, the other with semen. The environmental impossibilities that form the remade city represent an expression of Gothic excess. While death and violence erupt in the streets, New Paris is defined by its vivification, by an abundance of animation. As Miéville speculates in the afterword, 'the war- and dream-ruined city must teem' (p. 179) with the fantastical. While Surrealism may touch upon the Gothic's negative aesthetics, particularly the uncanny, it also laces it with an emphasis on the strangely wondrous.

Gothic anxiety is frequently evoked in response to the threat or collapse of boundaries. New Paris' manifs exist at the intersection of a host of collapsing boundaries for they are the corporeal impossibilities of the objects they depict, not merely the artistic depictions themselves. The city's surreal ecosystem brings into tension distinctions between nature and culture, idea and object, waking and dream states, psychology and ecology. In Miéville's new Urban Gothic, the rigid monumental structures of Victorian urban Gothic give way to surrealist fluidity and hybridity. Rather than the oppression of the static, the threat of the hidden, we are confronted with unresolved conceptual tensions and the visual disorientation of the polymorphous. This ontology of hybridity and uncertainty is best represented by the exquisite corpse, the manif that accompanies Thibaut and Sam. The exquisite corpse takes its name from a Surrealist 'game' invented by André Breton, Yves Tanguy, Jacques Prévert and Marcel Duchamp. Discarding rational thought and logical coherence through adopting a collaborative approach, each artist would add to an image on a folded piece of paper, the overall portrayal of which was not revealed until the end. As an exercise in Frankenstein monstrosity, New Paris is itself an exquisite corpse. Its hybridity fuses together fragments of the history of occupied Paris, a Surrealist exercise in reimagining the French capital that appeared in a 1933 edition of the journal, *Le Surréalisme au service de la révolution* (Miéville, pp. 192–193), and a rich assemblage of cultural references drawn from the history of Surrealism, Symbolism, Decadence and early twentieth-century occultism.[3]

The Gothic and Surrealism share an oppositional stance founded on the transgressive power and potential of the imagination. The emergence of Gothic literature can be understood as a response to Enlightenment rationality, one that flourished in the revolutions of the late eighteenth century and reinvented itself during the process of nineteenth-century urbanisation and industrialisation. Its deliberate emphasis on 'objects and practices that are constructed as negative, irrational, immoral and fantastic' (Botting 1996, p. 2) similarly informed the Surrealists' revolution against notions of the 'real'. Through tapping into and attempting to capture the feeling of interior realities, Surrealism sought to challenge the traditional view that art should focus on the representation and replication of external objectivities. In seeking to portray a 'psychic reality' in which 'the mimetic recording of objectivity is subverted in favour of … a state of mind rather than to a physical state' (Levy, p. 9), their art became sites of dreamlike disjunction. By placing ordinary images and objects into strange and de-familiarising juxtapositions the Surrealists demonstrated the liberating power of the imagination to challenge expectation and to disturb 'normality'.

Surrealism's combative stance, actualised in New Paris' conflicts between manifs, Nazis and demons, was evident from the movement's first manifesto in 1924. André Breton declared 'If the depths of our mind contain within it strange forces capable of augmenting those on the surface, or of waging a victorious battle against them, there is every reason to seize them'; in doing so dream and reality will form 'a kind of absolute reality, a surreality' (Harrison and Wood 1992, pp. 434 and 436). That the imagination was to be understood not just as a means of escape from reason and reality but a way of contesting them was made even more explicit in 'Towards a Free Revolutionary Art' (1938), a collaboration between Breton, Diego Rivera and Leon Trostsky. Challenging the repressive nature of totalitarian regimes, it declared 'true art is unable *not* to be revolutionary, *not* to aspire to a complete and radical reconstruction of society ... We believe that the supreme task of art in our epoch is to take part actively and consciously in the preparation of the revolution.' (Harrison and Wood, pp. 527–528).[4]

Some of the French Surrealists' identified revolutionary forebears included those with Gothic genealogies. Despite their supposed rejection of novel reading, Louis Aragon commented that his Surrealist companions were reading Matthew Lewis' *The Monk* (Aragon 1994, p. xii). They also saw themselves as continuing the equally provocative work of Isidore Ducasse, the Comte de Lautréamont, and the Marquis de Sade, both of whom were enshrined in the Marseilles tarot deck created by the Surrealists (Miéville, pp. 198–199). Given their interest in the unconscious, dreams and psychoanalytical techniques aimed at freeing the imagination, Freud was also an important figure for the Surrealists and openly mentioned in their first manifesto. He forms a significant lynchpin between Surrealism and the Gothic. Inspired to write 'The Uncanny' (1919) after reading E.A. Hoffmann's Gothic tale, 'The Sandman' (Smith and Wallace 2001, p. 4), Freud's essay has since become 'a foundational text for Gothic criticism' (Ridenhour 2013, p. 16). This is especially pertinent to the issues considered in this chapter, for it places heavy emphasis on psychological interpretations in which 'the physical spaces of the Gothic function ... as metaphors for the psychologies of its protagonists' (Ridenhour 2013, pp. 16–17).

Beneath the urban Gothic aspects, Miéville's novella is permeated with a political-aesthetic struggle that echoes that of the Surrealist manifestos. In historical reality, the Main à plume served as the rump of the French Surrealist movement in occupied Paris once its leaders had escaped. Their main role was as propagandists for the Surrealist cause (Walter and Vernay 2008), but in New Paris Miéville reimagines them more actively as 'soldiers of the unconscious' armed with Surrealist techniques that operate akin to magic (pp. 17 and 71). Although their motives remain unclear, the manifs' predisposition towards fighting demons and Nazis also positions them (and the Surrealist imagination which they articulate) as a form of moral resistance. ('Manif', an abbreviation of 'manifestation' in the novella, also means 'demo' or 'protest' in French.) Jack **Parsons** is initially disillusioned when he meets the pre-exile Surrealist leadership in Marseilles. Struck by the impotence of their games, he clearly promotes the view that resistance requires combat. Through his Golem magic the occultist seeks to make the inanimate animate, weaponising ideas so that they can have an effect in the fight against the Nazis.[5] This is best seen in the description of the S-blast as 'a city-wide outrushing, an explosion ... megaton imaginary, of random and of dreams ... a burning blast of unconscious'. (pp. 138–139). Surrealism's symbolic triumph over Fascism is represented by Thibaut's destruction of a Nazi manif, an animated Arno Breker sculpture, and the defeat of Fall Rot, an equally crude symbol of war in the form of 'a centaur of tank and great man-shape ... festooned with German flags'. (p. 152). Yet, indicative of the need to actively commit to the liberation of the imagination, when Thibaut finds himself at a liminal space on the boundaries of Paris, a place where he 'can go almost anywhere', he decides to return 'to the arrondissement he has known since he was a child. Where there is still a fight' (p. 168). It is Thibaut's commitment to the conflict that holds him in New Paris.

The story's final confrontation is less about politics and more about art and environmental aesthetics. The threatened dystopia is not that of the monstrously distorted, war-torn city. Architectural ruin, ecological transformation, the eruption of the grotesque and violent death, all these Gothic elements are to be preferred to the sanitised, empty city envisioned by the Nazi's most fearsome manif, the 'little blank-faced nonentity' (p. 164) of Hitler as a young artist. Its limited talent causes its gaze to depopulate the streets and replace them with perfected renditions of pretty, vacant buildings. Changing everything it looks upon, unable to rise above its own insipid vision, it threatens to make a new New Paris, one devoid of life and depth. Unlike the true art of Surrealism which probed interior mindscapes, the Hitler manif can only generate surfaces, for 'Behind the new façade its stare invokes, the rot of war remains' (p. 163). In this transformed city, a victory for Thibaut is aesthetic, achieved when the buildings 'remember their cracks' and return 'to ruination ... scarred with the stuff of history, again' (p. 166).

While Miéville's New Paris may echo the Gothic tropes evident in Thomson's poem, it also marks a significant move away from underlying ideas that developed in and came to pervade the Victorian urban Gothic tradition. Unlike Victorian Gothic horrors that remain largely concealed in London's vastness, in the new Urban Gothic secrets and individual interiorities are transformed into citywide landscapes; no longer hidden or repressed but exposed, manifested, unleashed. Whereas the nineteenth-century urban Gothic dwelt on labyrinthine complexity, architectural weight and the claustrophobic oppressiveness of brick and humanity, the new Urban Gothic speaks to millennial anxieties where horrors are made deliberately visible and where seeming solidity can be reduced to ruin in an instant. As a frequent site of terror(ism), a place where artistic and ideological manifestos can erupt into sudden violence in the streets, New Paris speaks to a contemporary urban mentality in which we live in a perpetual state of battle, daily traversing inescapable geographies of anxiety. Confronted by distorted fundamentalist ideologies and a resurgent right-wing popularism, Miéville's novella serves as a reminder of the struggle to free, or maintain the freedom, of the imagination in historical moments which have seen a wilful narrowing of perspective. If Thomson's poem was a dark paean to the melancholic pessimism of the Victorian city dweller, the new urban Gothic of Miéville's New Paris can be read as a manifesto championing the value of the grotesque artistic imaginary amidst the shifts of contemporary politics. Indeed, Miéville's text closes with an almost Wellsian message for the future: 'Perhaps some understanding of the nature of the manifs of New Paris, of the source and power of art and manifestations, may be of some help to us, in times to come' (p. 182).

However, given this book's concerns with the Anthropocene, we cannot conclude with such an unproblematic appreciation of the fantastical imagination. While the liberating potential to imagine alterity and otherness is to be valued and embraced, this is tempered by the recognition of an unspoken arrogance in the Surrealists' desire to transform the world. In their revolutionary promotion of the surreal over the real, the Surrealists seem to reflect a more imaginative variant of an apparently broader human impulse to destroy and remake our realities. The imaginative grotesquery of Miéville's new Urban Gothic encourages us to read his anarchic disruption of environmental 'normality' as a timely reminder of our need to both confront and adjust to the consequences of the Anthropocene. Yet in portraying this act of jarring distortion there is an indulgence in the transformation and overwriting of known places and histories of wartime Paris. The endless detonation of the S-bomb serves as his central metaphor for this imaginative, destructive urge. While Miéville's novella can be read as a provocative exercise in constructing notions of urban environmental alterity, the fantastical mode through which his messages are conveyed are, by their nature, rooted in a revelling in transfigured worlds. In the Anthropocene, it is not our Gothic cities that are inescapable; like the dwellers of the City of Dreadful Night, like Miéville's Thibaut, it is our own incarcerating inclinations.

Notes

1. For more on the French Gothic see A. Horner (2002), *European Gothic–-A Spirited Exchange 1760–1960* (Manchester: Manchester University Press), pp. 230–251.
2. The Gothic has shown itself capable of moving into both later periods of history and literature. For Gothic studies of modernist literature see Riquelme (2008), Walker (2002), and Smith and Wallace (2001).
3. Paris had already found earlier Surrealist expression in Louis Aragon's *Paris Peasant*. The occult nature of Nazi-occupied Paris was described in Yonnet's *Enchantements sur Paris* (1954; reprinted as *Paris Noir: The Secret History of a City*, 2006). For more on Surrealism's links to the occult see Bauduin (2014).
4. The second Surrealist manifesto (1929) had shifted the movement more directly towards Marxist revolution. Many Surrealists joined but had problematic relationships with the communist party.
5. The Golem of Prague, a creature of Jewish folklore, provides Miéville with a folkloric forerunner for his conceit of manifs as animated and sentient matter. Yet, much like *Frankenstein*, the Golem is a cautionary tale about the hubris of its creator, Rabbi Loew, who, along with creating a protector for Prague's Jewish ghetto, also sought knowledge and power over life and death. Echoing concerns about the Anthropocene, in the most famous version of this legend this product of human arrogance wreak havoc upon the very people and environment it was meant to protect. See Dekel and Gantt Gurley (2013).

Bibliography

Aragon, L. 1994. *Paris Peasant*. Boston: Exact Change.

Bauduin, T.M. 2014. *Surrealism and the Occult: Occultism and Western Esotericism in the work and movement of Andre Breton*. Amsterdam: Amsterdam University Press.

Botting, F. 1996. *Gothic*. London: Routledge.

Cavallaro, D. 2002. *The Gothic Vision: Three Centuries of Horror, Terror and Fear.* London: Continuum.

Dekel, E., and D. Gantt Gurley. 2013. How the Golem Came to Prague. *The Jewish Quarterly Review* 103 (2): 241–258.

Harrison, C., and P. Wood (eds.). 1992. *Art in Theory 1900–1990: An Anthology of Changing Ideas*. Oxford: Blackwells.

Heholt, R. 2016. Afterword: Affective Gothic Landscapes. In *Haunted Landscapes Super-Nature and the Environment*, ed. R. Heholt and N. Downing, 227–241. London: Rowman and Littlefield.

Horner, A. 2002. A Detour of Filthiness: French Fiction and Djuna Barnes's Nightwood. In *European Gothic—A Spirited Exchange 1760–1960*, 230–251. Manchester: Manchester University Press.

Kroger, L. 2013. Panic, Paranoia and Pathos: Ecocriticism in the Eighteenth-Century Gothic Novel. In *Ecogothic*, ed. A. Smith and W. Hughes, 15–27. Manchester: Manchester University Press.

Levy, S. (ed.). 1997. *Surrealism—Surrealist Visuality*. Keele: Keele University Press.

Mellor, L. 2011. *Reading the Ruins: Modernism, Bombsites and British Culture*. Cambridge: Cambridge University Press.

Miéville, C. 2016. *The Last Days of New Paris*. London: Picador.

Mighall, M. 2003. *A Geography of Victorian Gothic Fiction: Mapping History's Nightmares*. Oxford: Oxford University Press.

Phillips, L., and A. Witchard (eds.). 2012. *London Gothic: Place, Space, and the Gothic Imagination*. London: Continuum.

Ridenhour, J. 2013. *In Darkest London: The Gothic Cityscape in Victorian Literature*. Lanham: Scarecrow Press.

Riquelme, J.P. (ed.). 2008. *Gothic and Modernism: Essaying Dark Literary Modernity*. Baltimore: John Hopkin University Press.

Smith, A., and J. Wallace (eds.). 2001. *Gothic Modernisms*. Basingstoke: Palgrave.

Stiegler, B. 2012. *Uncontrollable Societies of Disaffected Individuals: Disbelief and Discredit*, vol. 2. Cambridge: Polity Press.

Thomson, J. 1993. *The City of Dreadful Night*. Edinburgh: Canongate Classics.

Walter, R., and A. Vernay. 2008. *La Main à plume: Anthologie du surréalisme sous l'Occupation*. Paris: Editions Syllepse.

Walker, R. 2002. Blooming Corpses: Burying the Literary Corpus in the Modern. *Gothic Studies* 4 (1): 1–13.

Wasson, S. 2010. *Urban Gothic of the Second World War—Dark London*. Basingstoke: Palgrave.
Yonnet, J. 2006. *Paris Noir: The Secret History of a City*, trans. Christine Donougher. Sawtry: Dedalus.

‘Things Are Not as They Seem’: Colonialism, Capitalism and Neo-Victorian London in *The Order: 1886*

Michael Fuchs

According to Fred Botting, transgression lies at the heart of the Gothic mode. Transgression, Botting has explained, ‘serves to reinforce or underline’ the ‘value and necessity’ of ‘social and aesthetic limits’ (1996, p. 5). The Gothic hence addresses taboos, features liminal beings and/or occurrences and depicts grotesque bodies, thereby testing social norms. These elements are part and parcel of the Gothic’s aesthetic principles, which are constantly repeated in individual Gothic texts. Indeed, one might even argue that the Gothic ‘is profoundly concerned with its own past, … familiar images and narrative structures’ (Spooner 2006, p. 10). As a result, any study of the Gothic is, in a way, ‘a study of repetitions’ (Lloyd-Smith 2005, p. 1). However, no repetition can perfectly re-create that which it repeats. As such, repetition always implies ‘difference … by way of disguise’ (Deleuze [1968] 1994, p. 17). In terms of genre works, this idea suggests that writers, filmmakers and other kinds of producers of cultural artefacts ‘imitate, learn from, and modify the work of their predecessors’ (Lloyd-Smith 2005, p. 1). Accordingly, the Gothic is rooted in

M. Fuchs (✉)
Department of English & American Studies, University of Oldenburg, Oldenburg, Germany
e-mail: michael.fuchs@uol.de

H.-G. Millette and R. Heholt (eds.),
The New Urban Gothic, Palgrave Gothic,
https://doi.org/10.1007/978-3-030-43777-0_3

the past; it builds on traditions, but at the same time, it is anchored in the present moment, reflecting on contemporary questions, which opens the door to innovation; it is a 'historicised practice which is durable yet transposable' (Jones 2015, p. 28).

The narrative-based videogame *The Order: 1886* (Ready at Dawn and SCE Santa Monica Studio 2015) exemplifies the operating principles of these processes, as it allows players to virtually travel to the birthplace of the Urban Gothic, Victorian London. To be sure, while the Gothic mode emerged in the eighteenth century as a response to the Enlightenment and its discourses of rationality and objectivity, the Urban Gothic developed in—and from the particular realities of—Victorian London. As the 'population of London soared from one million to over six million' in the course of the nineteenth century (Jackson 2014, p. 2), Londoners became increasingly 'fascinated and intermittently horrified by their developing urbanization' (Humphreys 2002, p. 602). To achieve its uncanny effects (and affects), the Urban Gothic of the nineteenth century fed on the paradoxes, ambiguities and contradictions embodied by the centre of the British Empire.

Visually, *The Order: 1886* draws on the '[i]mages of gaslit' and 'fog-enshrouded streets' (Ridenhour 2013, p. vii), which have become characteristic of imaginations of nineteenth-century London. However, the videogame does not simply repeat the images and narrative patterns typical of the Urban Gothic. As a Neo-Victorian cultural artefact, *The Order: 1886* enquires into 'act[s] of (re)interpretation, (re)discovery and (re)vision' in a self-reflexive manner (Heilmann and Llewellyn 2010, p. 4; original in italics). In so doing, the videogame exposes continuities between the past and the present by deploying Victorian London to reflect on the present moment. Indeed, whereas Rosario Arias has suggested that Neo-Victorianism uncovers the 'encrypted symptoms and traces of … physical and spiritual "sickness"' haunting the Victorians (2010, p. 136), the videogame's journey back in time allows the game to reveal 'our role as heirs of [a] continuous historical process' (Krueger 2002, p. xi), which is defined by exploitative practices under the banner of capitalist progress. In particular, *The Order: 1886* employs its medium specificities to great effect by supporting its critique of Empire through ludic means, as the videogame suggests that individuals have little to no agency in our world, thereby rendering them incapable of effecting change.

Performing the Past in the Present

The Order: 1886 exploits Victorian London's Gothic qualities before players start the videogame proper, as the PlayStation 4's background image displays a panorama shot of the game's urban setting when selecting the game (Image 1).

In the image, grey tones, smoke emerging from chimneys and fog dominate the scenery. By tapping into the well-established visual repertoire of Gothic London, the PS4 menu screen sets up players' expectations for the videogame they are about to play. This strategy continues upon launching the game. Following the movie-like 'Sony Computer Entertainment presents / and Santa Monica Studio External presents' lines (Ready at Dawn and SCE Santa Monica Studio 2015), which are presented in white colouring against a black background, an animated background—again in greys—appears as the opening credits continue to roll. After a few seconds, players reach the game's main menu, which is superimposed over the animated grey background. Overall, the colour palette and the slow visual movements set a tone of ambiguity and mystery.

When starting the game proper, the player-character is struggling underwater, highlighted by frantic movements of the virtual camera and

Image 1 PlayStation 4 menu image for *The Order: 1886*

sounds of air bubbles. Someone wonders, 'You think he's still breathing?' (Ready at Dawn and SCE Santa Monica Studio 2015). A second voice responds, 'I dunno. Why don't we pull him out and have a look?' (Ready at Dawn and SCE Santa Monica Studio 2015). However, the two male characters decide to wait for a few more seconds, pull out the player-character, push him underwater once again and then pull him away from the water trough. As the two guards begin to drag the player-character around, the scene's setting begins to take shape—the player-character is in a dungeon, probably part of a prison. After a few moments, they have reached the player-character's cell and leave him there. Exhausted, he falls unconscious only for the two men to wake him up sometime later (possibly even one day—or a couple of days—later) and drag him back to the water trough. On the way to the torture chamber, players can finally catch a glimpse of the face of the character they control, and an insert introduces the title of the first chapter—'Once a Knight' (Ready at Dawn and SCE Santa Monica Studio 2015). Another insert follows, which clarifies that the player-character is located in the catacombs of Westminster on 20 November 1886. As one of the warders again starts pushing the player-character underwater, players are asked to press a button. After pushing the right button, a series of further button commands allows the player-character to liberate himself. (These 'quick-time events' make up one major part of players' control of the action, next to traditional third-person shooter commands.) The player-character makes his way through the building only to end up on a wall walk, surrounded by guards and knights carrying futuristic weaponry.

Apparently, the knights know the player-character well, as one of them addresses him by his first name, Galahad (an heir to Galahad's seat at King Arthur's Round Table, as players learn in the course of the game). The oldest knight notes, 'You have betrayed our Order. No one may escape the penalty' (Ready at Dawn and SCE Santa Monica Studio 2015). Galahad replies, 'You forget, Lord Chancellor. I'm as good as dead already' (Ready at Dawn and SCE Santa Monica Studio 2015), and jumps off the building into the cold river flowing below. After a fade to black, the game's title appears in white letters. Then, players are transported farther back into the past, to 12 October 1886, when Galahad was still known as 'Sir Galahad'.

This outline of what amounts to roughly the first twelve minutes of gameplay in *The Order: 1886* serves to introduce two important aspects of the videogame. First, the opening minutes suggest that players are effectively reborn as Galahad and transported into a steampunk version of Victorian London. Narratively, the players' first contact with Galahad evokes a birthing scenario, as he emerges from the water, tries to catch a breath and gasps for air. As Galahad escapes the watery claws of death (only to jump into the icy water of the Thames a few minutes later), he is resurrected as a new man—no longer a knight, as the insert implies, but something else (or 'always one'—players cannot be sure at this point). This (re-)birth as something other than oneself also applies to players. After all, players need a virtual stand-in to act in the virtual world on their behalf. The interfaces of the gamepad and the avatar, the players' in-world representative, transforms players' inputs into actions in the virtual domain. Crucially, players must learn to act in the virtual environment, to control their virtual bodies, to move and to navigate the world. Jonathan Boulter has hence suggested that the players' first encounter with their virtual stand-ins marks an existentialist moment akin to what Martin Heidegger referred to as 'thrownness'—a human being's coming-to-consciousness in the world ([1927] 2010, pp. 131–143; pp. 172–175). '[T]he human', Boulter has explained, 'is simply thrown into the world [...] with no obvious guidance' as to how 'to make sense of the experience' (2015, p. 18). In the moment players are symbolically reborn as Galahad, their human bodies become interlinked with the console the game is running on, the gamepad in their hands and the player-character they control. In this way, players transcend the limits of their corporeal selves and perform actions in virtual worlds, as they become more-than-human.

As players phenomenologically merge with Galahad, the differences between the Victorian past imagined in the game and the present of play disappear. Indeed, videogame scholar Jesper Juul has argued that playing 'has a basic sense of happening *now*', since '[p]ressing a key influences the game world, which then logically (and intuitively) has to be happening in the same now' (2004, p. 134; italics in original). Whereas Juul has thus diagnosed 'an inherent conflict between the now of the interaction and the past or "prior" of the narrative' (2001, para. 50), this merging of different temporal layers contributes to the Gothic quality of *The Order: 1886*. As much as players are caught between virtual London and the material world outside, players (and their player-characters) are located between temporal dimensions, as well. In fact, the profusion of temporal layers in the opening minutes suggests that the events of 20 November 1886 will occur no matter the players' actions. Thus, the present moment of gameplay (narratively set in the past) is practically voided of meaning, as players come to understand that their virtual doubles are trapped in a pre-determined narrative that will inevitably unfold in a very specific manner.

On another level, the convergence of the imagined past and the present of playing literalizes the ways in which *The Order: 1886* performs the Gothic's past in the present (and anticipates its future repetition). The narrative structure of the tale as framed in the game's start-up spotlights this process, as the opening in late November and the subsequent travel back in time to mid-October implies that players will discover how Galahad not only ended up in the dungeon, but also why he and his apparent friends no longer see eye to eye. In so doing, the opening minutes draw on one of the most important characteristics of the Urban Gothic according to Kathleen Spencer, 'the effectual thematization of the mystery plot and the omnipresent figure of the detective, that characteristically Victorian discoverer of secrets' (1987, p. 92). Whereas earlier Gothic tales often undermine the God-like power of these men of logic, Galahad does, in fact, uncover the dark secrets lurking in the darkness; however, he can do little with this knowledge, as he is trapped in a rather powerless position. *The Order: 1886* thus self-reflexively engages with the Gothic's past–but in order to critically scrutinize the Gothic's past, it must be resurrected for the differences to become manifest. Along the way, the Gothic adapts to its changing socio-economic environment, as the videogame reimagines both the Gothic and the Gothic's past.

Moving Through Victorian London

One could make a similar argument about media, more generally. In their classic study ***Remediation***, Jay David Bolter and Richard Grusin conclude that media always *re*mediate—'*all* mediation is remediation', since 'each medium … responds to, redeploys, competes with, and reforms other media' (1999, p. 55; italics in original). Digital media, Bolter and Grusin continue, 'can be more aggressive in [their] remediation', as they 'refashion the older … media entirely, while still marking the presence of the older media' (1999, p. 46). The resultant media text 'becomes a mosaic in which we are simultaneously aware of the individual pieces and their new … setting' (1999, p. 47). A videogame such as *The Order: 1886* reimagines Gothic London, as established in Victorian cultural artefacts and remediated and adapted in film, television and other media. Significantly, this remediation necessarily adds media specificities to the mix, which influence our understanding of the cultural artefact in question.

In his essay 'Allegories of Playing', videogame scholar Espen Aarseth states that '[t]he defining element in computer games is spatiality', since videogames 'are essentially concerned with spatial representation and negotiation' (2001, p. 154). 'Negotiation' is the key word here, for whereas with videogames' representation of space responds to other media, in particular film, they are centrally concerned with movement through space, and, as such, with offering players spatial experiences. In particular, videogames set in cities usually seek to remediate the real world urban experience, which is inherently multimodal and multisensory.

When the action of *The Order: 1886* truly begins (i.e. 12 October 1886), players find Galahad surveying the city from a high building located somewhere in Mayfair. As Galahad walks out onto a platform, familiar sights such as Big Ben come into view (Image 2).

However, divergences from historical reality become manifest, as well: An imposing and phallic tower occupies the centre of the screen, surrounded by zeppelins hovering in the cloudy sky. These towers, as players learn in the course of the game, serve as bases for the airships, which project the realities of twenty-first-century surveillance culture back onto the late nineteenth century. This mixing of historical reality with present-day phenomena and fantastic elements draws players' attention to the fact that Galahad uses a mobile long-distance communicator, which

Image 2 Surveilling Victorian London in *The Order: 1886*

seems misplaced in the late nineteenth century, but clearly evokes a mobile phone. A few seconds of narrative exposition (without any real action) confirm that *The Order: 1886* taps into a tried-and-true steampunk formula by combining historical details with a pastiche of various images of Victoriana and adding 'fantastic technologies that might not have existed a hundred years ago' (Hantke 1999, p. 247).

Beyond establishing the virtual world's steampunk character, the opening view of the city calls to mind Michel de Certeau's elaborations on looking down at New York City from the World Trade Centre:

> When one goes up there, he leaves behind the mass that carries off and mixes up in itself any identity of authors or spectators. ... It transforms the bewitching world by which one was 'possessed' into a text that lies before one's eyes. It allows one to read it, to be a solar Eye, looking down like a god. ([1980] 1984, p. 92)

The view from the top allows Galahad to spot the United India House, where his companions Igraine and Lafayette wait for him. Once Galahad has reached the street level, players begin to discover a more 'elementary form of ... the city' defined by bodily experience (de Certeau [1980] 1984, p. 93). As Galahad walks through the streets of Mayfair, he can smell fresh meat (synaesthetically transformed into verbal remarks and visuals), he can hear the coughing of an apparently sick man and he can see advertisements virtually everywhere. Once the sun has set, electric lamps start buzzing all around him. Apparently, in *The Order: 1886*, the Victorian age is 'a textual construct open to manipulation and modification', as the past becomes 'strikingly and conspicuously *unreal*' (Hantke 1999, p. 248; italics in original).

The violent encounters in the city of London emphasize the imagined past's fantastic character. After clearing the United India Company's premises of rebels, Galahad and company proceed to London's underground system, where they find more troublemakers. Galahad only comes to the understanding that '[o]ne of [their] escapees is a Half-Breed', before discovering that there is 'more than one' (Ready at Dawn and SCE Santa Monica Studio 2015) immediately after that. These half-breeds are part human, part animal who can change their shape at will. The half-breeds in the underground, in particular, remind the reader of werewolves (in fact, the on-screen display refers to them as 'lycans')—minus the full-moon madness (Image 3).

Image 3 The lycans are among the fantastic elements The Order: 1886 adds to nineteenth-century London

Galahad quickly disposes of two creatures, but the third one escapes to the surface. As the monster runs through the streets of Mayfair, people seem afraid, but not particularly surprised to see the beast. Evidently, even places rather close to Buckingham Palace cannot guarantee of safety. The fact that fantastical events even happen in these wealthy districts of London undermines the very notion of what it is to be safe in the Victorian city. Whereas early 'urban gothic tales [...] derive [...] their emotional power from the contrast between the materialist beliefs of most of the characters and the supernatural marvels or horrors happening before their eyes' (Spencer 1987, p. 91), in *The Order*, fantastic creatures are part of everyday life in the city.

Significantly, the Commissioner opines that because of 'the Rebellion, ... this "Jack" and the Ripper murders', Whitechapel 'has become the epicentre of [London's] recent conundrums' (Ready at Dawn and SCE Santa Monica Studio 2015). By corroborating the connection between fantastical creatures (i.e. 'the Rebellion') and the historical (1888) Whitechapel Murders, *The Order: 1886* reveals the role Whitechapel has had in imagining London: The East End functions as the binary opposite to the (not only) wealthy, (not necessarily) beautiful and (not always) bright centre of London. As Jamieson Ridenhour tellingly explains in his study of representations of Gothic London in Victorian literature, 'the Gothic districts of London replace the foreign settings and villains of the Radcliffean Gothic', as 'the East End allows the teleological narrative of the Gothic to be told in the modern city' (2013, p. 52), which perfectly explains the East End's symbolism in this Gothic game.

Lycans, Vampires and (Post-)Colonialism

As Galahad navigates through the 'infernal city' (Ready at Dawn and SCE Santa Monica Studio 2015) of London, he proceeds deeper into its run-down, darker areas. Along the way, he encounters other fantastical creatures, such as vampires, and seems to come closer to uncovering who is behind the rebellion and who pulls the fantastical creatures' strings. All pieces of evidence point at a mysterious woman who turns out to be from India. She approaches Galahad in a Whitechapel tavern (which also functions as a brothel—both a cliché and a historical reality), where she reveals herself to be Devi Nayar, daughter of Lakshmi Bai, who is the actual leader of the rebellion. Galahad comes to understand that Lakshmi is 'the Rani of Jhansi', a 'famous warrior queen who took up arms against [the British] army in India' (Ready at Dawn and SCE Santa Monica Studio 2015). British 'generals reported [her] killed in Gwalior' a number of years ago (Ready at Dawn and SCE Santa Monica Studio 2015).

Here, the videogame again draws on historical reality. Lakshmibai was one of the leading figures in the Indian Rebellion of 1857, which eventually led to the dissolution of the East India Company and the reorganization of British administration in India. Lakshmibai died close to Gwalior during the Rebellion, although British forces never found her body. While the videogame's creative manipulation of history becomes evident in this case again, in her role as a fierce warrior stepping up to protect her people from outside forces, the Indian woman embodies the first stage of resistance against imperialism Edward Said has outlined—'primary resistance'—which entails 'literally fighting against outside intrusion' (1994, p. 252).

Lakshmi also embodies the second stage of resistance, 'ideological resistance', which implies the attempts 'to save and restore the sense and fact of community against all the pressures of the colonial system' (Davidson 1978, p. 155). Indeed, she does not only seek to stop the United India Company from further expanding westward, but wants to bring down the entire corporation. The Indian Rebellion was 'a restorative revolt ... against Company rule' (Robins 2012, pp. 194–195); likewise, Lakshmi's rebellion in the heart of London tries to undo the forces of multinational capitalism that the East India Company so successfully instituted. Indeed, Lakshmi tells Galahad that England is 'an empire of bootlickers, grovelling at the feet of the mighty United India Company' (Ready at Dawn and SCE Santa Monica Studio 2015).

Galahad does not want to believe that he is little more than an easily manipulatable and disposable pawn in a game of chess the Company plays. However, Lakshmi shows the knight that the boxes destined for America and the Caribbean are filled with vampires. 'With the protection of the Lycans, [Lord] Hastings and his vampire ilk are spreading disease throughout your city' (Ready at Dawn and SCE Santa Monica Studio 2015), she explains. When Galahad protests that 'Hastings can't be one of them!', she remarks that Hastings 'prey[s] on downtrodden women' (Ready at Dawn and SCE Santa Monica Studio 2015) late at night. As the interconnections between The Whitechapel Murders, the proliferation of fantastical creatures in London and the rebellion dawn on Galahad, the symbolic significance of the vampires as representatives of the capitalist order become evident, as well: they feed on others in order to survive (or, rather, remain undead), exploiting them for personal gain.

In his *Critique of Political Economy*, Karl Marx claimed that 'capital is dead labour which, vampire-like, lives only by sucking living labour, and lives the more, the more labour it sucks' ([1859] 1990, p. 342). By drawing on the links between vampirism and capitalism, *The Order* suggests that the imperial centre 'literally devours its populace' (Witchard and Phillips 2010, p. 1) through its vampiric tools of exploitation. Revealingly, the vampires' hunting grounds are both in the faraway colonies and in London itself, thereby equating the lower class inhabitants of London's East End with the marginalized populations elsewhere—and the fanged beast that is capitalism victimizes all of them. Since capital 'feasts on the dead, and in doing so devours all life' (McBrien 2016, loc. 2424), the vampires' immortality combined with their sheer numbers and the operating principles of the capitalist endeavour predict a dire future for everyone involved.

When Galahad finally comes to comprehend that Lord Hastings is the infamous 'Jack the Ripper', perpetrator of the Whitechapel Murders, and planned to use the United India Company's global business entanglements to have his vampires colonize the entire planet, he is charged with treason and imprisoned, as the narrative returns players to the opening moments of the videogame. After his audacious jump into the icy water of the Thames, Galahad remains in a liminal state of being—neither quite dead nor quite alive—for close to two weeks. The first day he has (somewhat) regained control over his body, Galahad confronts his former confidant, Sir Lucan, the Lord Chancellor's adoptive son, who is a vampire. The scene ends in a boss battle. Severely injured, Lucan remarks that over the course of the centuries he has lived, that he has seen '[c]ivilizations born and destroyed by humanity's incessant greed' (Ready at Dawn and SCE Santa Monica Studio 2015). Vampires, on the other hand, primarily try to survive.

When the Lord Chancellor appears, Galahad requests that he reveal the truth about his adoptive son's being a vampire, but the Lord Chancellor refuses, saying, 'The revelation would shake our Order to its very foundations' (Ready at Dawn and SCE Santa Monica Studio 2015). The Lord Chancellor goes on to expel Galahad from the Order, for that is the only way for it to 'remain united' (Ready at Dawn and SCE Santa Monica Studio 2015). Galahad kills Lucan, and the screen fades to black. Throughout, players do not make any meaningful decisions; the videogame does not offer players any choices beyond in what order and in what ways to dispose of their foes. In this way, *The Order: 1886* exposes players' lack of control over the trajectory the narrative will take. Players are, as Lev Manovich observed, made 'to follow pre-programmed, objectively existing associations' (2001, p. 61), thereby steering their actions into very specific directions.

In the end, Galahad might have defeated Lucan, but Lucan was merely a tiny piece to a much larger puzzle that remains unresolved. In a final, brief, cinematic sequence, set in a rainy night, Galahad looks at the brightly lit United India House from a rooftop. The visual construction carries an ironic tone, for United India is staged as a beacon of light in the middle of the surrounding darkness. However, players know that United India is the source of many problems haunting virtual London. Ridenhour has argued that 'the threat of the Gothic is not that these "vestigial stains" exist, but that they cannot be contained by Southern Europe, or equatorial Africa, or the East End of London' (2013, p. 59). *The Order: 1886* turns the tables, as the monstrous threat actually spreads out from Mayfair, originates in the purportedly civilized and cultured part of London. In this way, the videogame exposes a shortcoming in Ridenhour's argument—the monstrous threat does not reside on the outside, trying to invade the untarnished and innocent inside that is London; instead, London may not necessarily be the monster itself, but its lair, providing shelter to the monstrous excesses of imperialism and capitalism.

Galahad—as an individual—can do little to fight the systemic issues that the imperial centre of London embodies.

Our Victorians and Empire in the Present Moment

'Over and over the tales of the urban gothic repeat this message: surface appearances cannot be trusted', Spencer has observed (1987, p. 92). 'The natural, reasonable, mundane world is only a thin film', she continues, 'covering a realm of horrors which at any moment might break through to attack the unsuspecting' (1987, p. 92). However, where Spencer opines that the city 'creates a hospitable space in the modern world for such monsters, a place where they can hide unsuspected' (1987, p. 92), in *The Order: 1886*, the monsters 'hide' in plain sight. Indeed, monsters control the Victorian metropolis to the point that the city itself becomes monstrous. While Peter Ackroyd, in his biography of London, diagnoses that the British capital became 'the climax, or the epitome, of all *previous* imperialist cities' (2001, p. 575; my emphasis) during the course of the nineteenth century, in *The Order: 1886*, it not only symbolizes the colonialist practices of the Victorian age, but it also epitomizes the capitalist excesses of today—and possibly the future.

The Gothic, Robert Mighall has noted, 'cannot be an essence, for what is Gothicized constantly changes. This depends on how each culture chooses to represent itself, and where it locates progress and its necessary antithesis' (1999, p. 286). Although this statement holds true, one cannot ignore the generic trappings that require repetition for a cultural artefact to be considered a Gothic text, in the first place. Players (but also viewers, readers and creators) have a tendency to constantly return to the same narrative patterns and ludic operations for the simple reason that we know them; they are reassuring and thus allow us—at least momentarily—to feel in control. Following John Rieder, these tales offer 'ideological fantasies', which are 'beliefs that we consciously disavow, recognizing them as untrue, but nonetheless support in practice' (2008, p. 30).

The Order: 1886 exposes Galahad's powerlessness in view of the capitalist system. Galahad might have eradicated an entire shipment of bloodsuckers (an act that might have momentarily stalled the exploitation of the New World) and he might have killed one of the most powerful vampires haunting London, but these actions will not have any large-scale effects on the fate of the world. After all, the vampires are agents of the United India Company, a thinly veiled stand-in for the East India

Company, which shaped global economic history: 'No stranger to stock market bubbles, eye-watering corruption and government bail-outs, the Company actually outstripped the excesses of the contemporary corporation by conquering nations and ruling over millions with its private army' (Robins 2012, p. xi). Crucially, the United India Company's power is concentrated in London with vampires as their imperial agents of that power and this represents the 'singular power of a *new city*' (Hardt and Negri 2000, p. 395; italics in original).

To be sure, the 'continued relevance of Empire … in one of the newest media', Suvik Mukherjee concludes in his book on postcolonialism in videogames, 'is one that is both worrying as well as, arguably, symptomatic of the ambiguity with which contemporary (particularly Western) society views the imperialist system' (2017, p. 30). Indeed, 'Empire', Michael Hardt and Antonio Negri have suggested, 'operates on all registers of social order' (2000, p. xv). Effectively, we are made to play along with Empire. And while numerous videogames require players to play their roles in promoting and perpetuating Empire (Patterson 2015; Fuchs et al. 2018), *The Order* confronts players with the lack of agency they have. In the end, if they want to finish the game, they have to play their role; they cannot stop the capitalist system embodied by the United India Company. In this way, the videogame suggests that, in some shape or form, every one of us participates in the capitalist processes characteristic of the neoliberal empire we not only inhabit (probably in a city) but are complicit in sustaining.

References

Aarseth, E. 2001. Allegories of Space: The Question of Spatiality in Computer Games. In *Cybertext Yearbook 2000*, ed. M. Eskelinen and R. Koskimaa, 152–171. Jyväskylä: University of Jyväskylä.

Ackroyd, P. 2001. *London: The Biography*. London: Vintage Books.

Arias, R. 2010. Haunted Places, Haunted Spaces: The Spectral Return of Victorian London in Neo-Victorian Fiction. In *Haunting and Spectrality in Neo-Victorian Fiction: Possessing the Past*, ed. R. Arias and P. Pulham, 133–156. Basingstoke: Palgrave Macmillan.

Bolter, J.D., and R. Grusin. 1999. *Remediation: Understanding New Media*. Cambridge, MA: MIT Press.

Botting, F. 1996. *Gothic*. London: Routledge.

Boulter, J. 2015. *Parables of the Posthuman: Digital Realities, Gaming, and the Player Experience*. Detroit: Wayne State University Press.

Davidson, B. 1978. *Africa in Modern History: The Search for a New Society*. London: Allen Lanes.
de Certeau, M. [1980] 1984. *The Practice of Everyday Life*, trans. S. F. Rendall. Berkeley: University of California Press.
Deleuze, G. [1968] 1994. *Difference and Repetition*, trans. P. Patton. New York: Columbia University Press.
Fuchs, M., V. Erat, and S. Rabitsch. 2018. Playing Serial Imperialists: The Failed Promises of BioWare's Video Game Adventures. *The Journal of Popular Culture* 51 (6): 1476–1499.
Hantke, S. 1999. Difference Engines and Other Infernal Devices: History According to Steampunk. *Extrapolation* 40 (3): 244–254.
Hardt, M., and A. Negri. 2000. *Empire*. Cambridge, MA: Harvard University Press.
Heidegger, M. [1927] 2010. *Being and Time*, trans. J. Stambaugh. Albany: State University of New York Press.
Heilmann, A., and M. Llewellyn. 2010. Neo-Victorianism and Post-Authenticity: On the Ethics and Aesthetics of Appropriation. In *Neo-Victorianism: The Victorians in the Twenty-First Century, 1999–2009*, ed. A. Heilmann and M. Llewellyn, 1–32. Basingstoke: Palgrave Macmillan.
Humphreys, A. 2002. Knowing the Victorian City: Writing and Representation. *Victorian Literature and Culture* 30 (2): 601–612.
Jackson, L. 2014. *Dirty Old London: The Victorian Fight against Filth*. New Haven, CT: Yale University Press.
Jones, T. 2015. *The Gothic and the Carnivalesque in American Culture*. Cardiff: University of Wales Press.
Juul, J. 2004. Introduction to Game Time. In *First Person: New Media as Story, Performance, and Game*, ed. P. Harrigan and N. Wardrip-Fruin, 131–142. Cambridge, MA: MIT Press.
———. 2001. Game Telling Stories? A Brief Note on Games and Narratives. *Game Studies*, 1.1, http://www.gamestudies.org/0101/juul-gts/. Accessed 15 April 2018.
Kohlke, M.-L., and C. Gutleben. 2012. The (Mis)Shapes of Neo-Victorian Gothic: Continuations, Adaptations, Transformations. In *Neo-Victorian Gothic: Horror, Violence and Degeneration in the Re-Imagined Nineteenth Century*, ed. M.-L. Kohlke and C. Gutleben, 1–48. Amsterdam: Rodopi.
Krueger, C.L. 2002. Introduction. In *Functions of Victorian Culture at the Present Time*, ed. C.L. Krueger, xi–xx. Athens: Ohio University Press.
Lloyd-Smith, A. 2005. *American Gothic Fiction: An Introduction*. New York: Continuum.
Manovich, L. 2001. *The Language of New Media*. Cambridge, MA: MIT Press.
Marx, K. [1859] 1990. *Capital: A Critique of Political Economy*, vol. I, trans. B. Fowkes. London: Penguin.

McBrien, J. 2016. Accumulating Extinction: Planetary Catastrophism in the Necrocene. In *Anthropocene or Capitalocene? Nature, History, and the Crisis of Capitalism*, ed. J.W. Moore. Oakland, CA: PM Press, Kindle edition.

Mighall, R. 1999. *A Geography of Victorian Fiction: Mapping History's Nightmares*. Oxford: Oxford University Press.

Mukherjee, S. 2017. *Videogames and Postcolonialism: Empire Plays Back*. Basingstoke: Palgrave Macmillan.

Patterson, C.B. 2015. Role-Playing the Multiculturalist Umpire: Loyalty and War in BioWare's *Mass Effect* Series. *Games and Culture* 10 (3): 207–228.

Ready at Dawn and SCE Santa Monica Studio. 2015. *The Order: 1886*. San Mateo, CA: Sony Computer Entertainment, PlayStation 4.

Ridenhour, J. 2013. *In Darkest London: The Gothic Cityscape in Victorian Literature*. Lanham, MD: Rowman & Littlefield.

Rieder, J. 2008. *Colonialism and the Emergence of Science Fiction*. Middletown, CT: Wesleyan University Press.

Robins, N. 2012. *The Corporation That Changed the World: How the East India Company Shaped the Modern Multinational*. London: Pluto Press.

Said, E.W. 1994. *Culture and Imperialism*. London: Vintage.

Spencer, K. 1987. Victorian Urban Gothic: The First Modern Fantastic Literature. In *Intersections: Fantasy and Science Fiction*, ed. G.E. Slusser and E.S. Rabkin, 87–96. Carbondale: Southern Illinois University Press.

Spooner, C. 2006. *Contemporary Gothic*. London: Reaktion Books.

Witchard, A., and L. Philips. 2010. Introduction. In *London Gothic: Place, Space and the Gothic Imagination*, ed. A. Witchard and L. Philips, 1–5. London: Continuum.

A Very Queer Black Country

R. M. Francis

Following Manuel Aguirre, this chapter perceives the Urban Gothic as a liminal landscape, where off-kilter figures pass through threshold experiences. Aguirre argues that the 'Gothic is fundamentally a *spatial* genre … pitting the familiar reality against another which is irreducible to it …. The (literal or figurative) crossing of the threshold is perhaps the prototypical deed in Gothic fiction, and versions of it include not just the hero's journey but also transformation, transgression, imprisonment, haunting, escape and pursuit' (2017, pp. 299–301). Narratives set in the Urban Gothic often focus on sexuality, otherness and radical transgression—queer characters in these landscapes embody this kind of Gothic liminality better than most. The horror and weird fiction of Joel Lane are examples of these landscapes and characters in flux, opposition and transition, and these texts make a feature of the queer and the queered.

Lane's short stories present the dark and often seedy underbelly of Birmingham and the Black Country as places of crime, depravity and the supernatural. His characters struggle with loss/lack and are often aimless. Ramsey Campbell notes the politic in Lane's work as well as a 'deep

R. M. Francis (✉)
University of Wolverhampton, Dudley, West Midlands, UK
e-mail: r.francis@wlv.ac.uk

H.-G. Millette and R. Heholt (eds.),
The New Urban Gothic, Palgrave Gothic,
https://doi.org/10.1007/978-3-030-43777-0_4

humanism and sympathy for characters that are often the excluded or oppressed' (Campbell 2016). Lane's landscapes are odd and out of joint post-industrial tableaux and ruined waste grounds, such that, as Peter Coleborn describes it, reading Lane is like 'reading a map of the tragedy of that city and its environs' (Coleborn 2016). Similarly, Chris Morgan refers to Lane as a 'great chronicler of local slum estates, crumbling tower blocks, derelict factories and dark canals' (Morgan 2016). This chapter will discuss several pieces from Lane's *The Lost District, and Other Stories* (2006) that narrate formative experiences within the decaying landscape of the Black Country, and from *Where Furnaces Burn*, his 2012 collection set in Wolverhampton. Focusing on Lane's choice of Urban Gothic locations in Black Country settings, this essay will explore how the natural and the unnatural are in an almost constant state of flux, in these post-industrial topoi. In the boundaries between the real and unreal, the dead and alive, the past and present, familiar and unfamiliar; in the constancy of the in-between, Lane's characters experience scenes of the strange as well as scenes of non-normative sexual encounters.

In both collections, Lane's settings are ones where urban decay and the weeds of semi-rural space wrestle for prominence. Lane writes with a sense of pride and nostalgia for the area's industrial heritage and working-class ethics. He also writes with a sense of despair towards the ruins, waste grounds and new enterprises that now replace these. He seems to have an infatuation with the contradictory meetings of the green and grey spaces. Using the local area, its history and its vernacular, Lane writes of liminal characters and experiences that read as Freud's *unheimlich*: as uncanny. The uncanny is a central tenet of the Gothic, precisely because it represents the return of that which should have remained dead and buried and it frightens us with what ought to be familiar. Lane achieves this with the landscapes he draws, in the main. These are all familiar to the protagonists—they are Black Country towns the characters either do not leave, cannot leave or do return to, yet they all harken back to their industrialist pasts. He occasionally does this with characters that die and reanimate or should be dead but still live, but in both cases the strength of the relationship between these uncanny spaces and subjects is that they both entice, while simultaneously repelling or disgusting the characters or reader. Such a paradoxical experience is further understood as sublime—a second central tenet of the Gothic—in that what draws us into, but also causes us, anxiety. Lane's cultural geography sets the theme for these texts: borderless, ghostly, uncanny and insecure narratives that entice and threaten but excite and invite in equal measure. In the in-between of the Black Country, Lane chooses to excavate queer or queered subjectivities as well as queer sexualities.

While important work has been done within LGBTQi communities in terms of combating prejudice and fighting for equality, the focus on the metropolitan and the liberal-gay world can be seen as a limiting and/or exclusive for those who occupy more parochial or marginal socio-geographical positions. The queer subject looks for space that is more fluid than these normalised liberal, metropolitan spaces, and as such the very notion of 'safe space' has often become dislocating for all those emerging with queer identities. With its marginal and liminal resonances, Lane exploits the post-industrial landscape in the centre of England, drawing it as an alternative non-homonormative urban space that counters the liberal, metropolitan gayborhoods of the big cities and the assimilationist suburbia of the leftist elite. The post-industrial landscape, like the queer body, is fluid and unboundaried. Located in the centre of England and known as the Cradle of the Industrial Revolution, the Black Country sits in the shadow of its bigger, more successful cousin, Birmingham. The Black Country is borderless—not even its population can agree where its boundaries are. It is not quite south, not quite north, a place that is difficult to map, where new meets old, where rural and urban mix and where an industrial heritage—a fundamental marker of the region's identity—has been replaced by enterprise zones and retail centres. Its towns—originally market towns—include, but are not limited to: Walsall, Wolverhampton, Dudley and West Bromwich. The industry of its towns sprung up along the coal seam that shot through the area—the place, was and still is, geographically and socio-politically on the limen. Joel Lane's short stories exploit the Black Country's post-industrial liminality and in this chapter, I discuss the ways in which, this new Urban Gothic topos offers a fertile time-space for queer subjects and experiences.

Heading into Lane's Gothic and Non-Normative in 'The Lost District'

In his story 'The Lost District' in the 2006 collection of the same name, Lane uses the area's threshold qualities to deliver post-deconstruction's and post-neoliberalism's nihilistic message of self-absorption and apathy. 'The Lost District' tells the story of an unnamed man, returning to the fictional Clayheath, a town he fondly remembers from childhood, where he encounters the ghosts in the dilapidated locale and the ghosts of his past. He reflects on a past girlfriend, Nicola, and memories from his teenage years—recollections that fuse with the ruined landscape and hamper his attempts to recuperate his lost town. The short-story begins: 'we were too obsessed with our needs and resentments to communicate. None of us knew what to say, what to feel, what to believe in. It didn't matter: nothing was going to change' (Lane 2006, p. 1). Lane's narrator is referring here to a general collective of disenfranchised people; it's inferred that these are the Generation Xers and early millenials of Clayheath. Such apathy and disaffectation queers and darkens the narrative. Clayheath is then drawn as a landscape that complements this apathy and disaffectation:

> All around us were raw traces of industry a hundred years old: canals just below road level, a brickworks wearing a loose scarf of smoke, black cast-iron railings and crudely worked flowers, walls studded with blue-green pieces of clinker from glass manufacture. By contrast, the houses themselves were coldly uniform: narrow grey terraces arranged in regular grids like the lines on a chessboard. The district seemed overcast, though the sky was dead white. (Lane 2006, p. 3)

Here, Lane draws a post-industrial landscape ruined by free-market economics and late-capitalism. The whole of the introduction sets up an Urban Gothic backdrop of resentment and unfulfillment with characters that have been reduced to a zombie state: 'I applied myself to the task of growing up in Thatcher's new Britain' the protagonist reports, 'Instead of becoming a journalist, I became an accountant. Somehow, my curiosity about the world had gone' (Lane 2006, p. 11). Lane's characters and culture are the undead of the Black Country, merely existing, parasitic, aimless.

Clayheath is drawn as border territory in this story. The rural and the organic fuse with the manmade and industrial, sometimes so much so it is difficult to separate these oppositions. An example of this is Lane's description of trees: 'silvery trunks slashed with rust' (Lane 2006, p. 2) and 'the line of poplar trees like huge railings' (Lane 2006, pp. 2–7). Fusing the urban with the rural; the organic with the machine, he illustrates the hybridity of the post-industrial at the millennium, where old and new merge. This mixing of spaces that tempt with their safe-unsafe/homely-unhomely aspects, draw the Black Country Urbanity as uncanny, but also as haunted. The people within Clayheath are odd, haunted, half-beings with eyes of 'hollows of darkness' (Lane 2006, p. 3) and 'you couldn't see their teeth when they spoke … Even their speech is unreal, as if the dead had been reanimated: it was like ordinary words in mouths that weren't really quite alive' (Lane 2006, p. 7). The gaps between towns were a mixture of rural and industrial features: forests, wasteground, factories, scrap yards, canals. Parts of the line ran close to the backyards of terraced houses, where clothes 'jittered on washing lines and blurred figures moved behind windows' (Lane 2006, p. 3). Lane's Black Country is uncomfortable, it's a place on the periphery that both threatens and entertains its reader with its haunting. The characters that populate Clayheath are also peripheral, they exist in a psychic borderland and this behaves as a psychic-loop for the story's main character, a loop that threatens his subjecthood.

What we discover as the story progresses is that the region is, in point of fact, invisible—it does not appear on the map: 'when I looked it up in my new A-Z, the district no longer existed' (Lane 2006, p. 11), says the protagonist. Instead, the area has been obliterated by the new: 'a jumble of housing blocks and prefabricated industrial units sprawled between the pitted roads' and '[a] new expressway cut across the end of the high street before circling around a giant plastic-fronted garden centre' (Lane 2006, p. 11). The selfhood and heritage of the town has been, by turns, obscured and obliterated by the modern.

Like the landscape, Lane's characters are both abject and liminal as well. The main character has a sexual encounter, allowing the reader access to his sexualised subjecthood, but this encounter is, itself, uncanny. When he first sees the object of his desire, she is faceless and doubled—another Gothic trope: 'below me, the girl was standing. I could see her upturned face, almost featureless at this height. A sudden vertigo snapped my eyes out of focus and I could see two of her, no less alone for it' (Lane 2006, p. 2). When they engage in sex, it too is foreboding: 'she showed me how to give her pleasure with my fingers and I felt less guilty … I had an unmistakable sense of being watched' (Lane 2006, p. 5). The narrator's paranoia at being watched and his novice status of not knowing how to pleasure his lover, suggest a certain ambivalence. Guilt, fear, lust and desire are all in flux here—the experience is neither anchored nor safe. Further, the sex sits off kilter and almost out of time and draws him back into his foundational memories of childhood sexual discovery. A memory whose language of sexual naiveté is both Gothic and disturbing:

> the little car park with its row of disused garages where, when I was eleven, the twin girls who lived up the road led me through a broken wall into a sealed-off alley where they showed me their vaginas [...] folds of pale skin reminded me of bacon left too long in the fridge. I had to expose myself too; it was probably the first time I had an erection, which I took to be somehow an effect of the cold. (Lane 2006, p. 6)

There is the use of the Gothic double with the twins, uncanny in the urban threshold space that is close to home but not homely, and the disturbing imagery of the girls' genitals—all prose that can be considered sublime.

This is a short story that deals with the unpacking of a non-normative sexual experience while in the process of returning—returning 'home' to find one's selfhood in an abject and derelict landscape. Sex and the identity politics run in tandem with the natural and the mechanical; the present and the past. In the in-betweenness of a disappearing and disappeared Clayheath, everything seems both real and unreal. Sexual awakening, in this example of Lane's work, is an act of becoming simultaneous to a wrestling with the past. The story's action, landscape and characters are in flux and subjects of doubling—an Urban Gothic where bodies, time and space no longer belong to their rightful owners, or sit in their rightful, logical place. This is not a romantic sketch of heteronormative love, rather it is a tale of the weird in desire experienced by subjects who feel othered, out of joint and disturbed.

Heading Down a Queer Lane in 'Scratch'

Lane uses the post-industrial landscape as a setting where non-normative and/or queer sexualities can come to fruition. His Black Country is more than a stage for the action, it becomes something larger—a catalyst or even a cause of queerness. In the 'Scratch'—the shape-shifting tale of a man–cat hybrid—the quasi-urban housing estates are home to non-normative sex acts, non-normative articulations of desire and illusions to bestiality. The gender-fluid protagonist of 'Still Water' falls somewhere between human and non-human and uses the ruins of the post-industry town as part of its reproductive process. Both sadomasochism and necrophilia are central to 'Wake up in Moloch', in which the ghost of the machines that gave the region its industrial power, now demands appeasement. And in 'My Stone Desire' a foundational queer experience is set at an old railway bridge where the organic and the manmade in the Urban produce gratification via putrefaction. As in 'The Lost District', Lane uses the post-industrial landscape of the Black Country as the weird setting where these Othered sexualities are explored or drawn out. Indeed, the Urban Gothic landscapes of Lane's Black Country are more than scenography in these stories, they are characters and catalysts for the weird and the queer.

Here, 'Queer' will be used in all three of its connotations: (1) in its original form, meaning 'strange' or 'peculiar'; (2) as an umbrella term for non-normative sexual behaviours and subjectivities and (3) as a de-individualised and politicised identity bracket for a sexual minorities. Queer Theory suggests that gender, sex, preference and the inherent sociopolitical issues that tie-up with these, all form part of a Queer cooking pot. There is no finite reason for these particular lifestyles, choices or wants, rather sexual desires and sexuality remains fluid and unmappable in this paradigm. Stein explains that 'queer' refers to any 'non-normative sexuality which transcends the binary distinction homosexual | heterosexual to include all who feel disenfranchised by dominant sexual norms' (Stein 1992, p. 838), and this is how it is, predominantly, used here. Considering queer subjectivities in this way is similar to how Gothic Theory considers Monsters and the othered—as threshold forces. Thus, as the monster and the Other is always already within us not external to us, Lane's short-fictions remind us that queer subjectivities and disenfranchisements are also always already within us. Lane's stories all sit within the threshold of these forces and recount formative rites of passage and othered sexualities. A particularly good example of this is Lane's story 'Scratch', from his *The Lost District, and Other Stories* (2006).

'Scratch' is set in Oldbury, a small town between Dudley and Birmingham. Lane depicts this as a landscape in flux, where the familiar and unfamiliar, domestic and wild, urban and rural and safe and unsafe space are all vying for prominence. The tale is told in the first-person, and our narrator is queer (as in, peculiar of character) who is ill at ease with his post-industrial life. He has an affinity with his pet cat, Sara, whom he sees as apart from the human species—'I don't think she ever forgave the human race for taking her identity away.... People say there's no such thing as a domestic cat, and it's true...whatever you feed them, they still hunt. When they bring you something they've killed, it's not a gift. It's a lesson' (Lane 2006, p. 31). He has a sublime sense of detachment with the animal, 'A cat's world is full of territories, friends and enemies, safe roads and dangerous roads... without her I felt – not alone, because I always felt alone, but like I wasn't all there' (Lane 2006, p. 31). He identifies with the cat, recognising himself in its half-way house of existence. He sees the cat in Freudian terms; as a lack or as a potential fear of castration. The cat is both domestic (*heimlich*) and strange (*unheimlich*). This is in keeping with Freud's ideas of ambivalence in the Uncanny and what Hugh Haughton commented on as being 'the sublime territory of unfamiliarity itself' and 'a paradoxical mark of modernity' (Haughton 2003, p. xlix).

The narrator is not only psychically attached to his cat, as the story progresses he becomes more cat and less human. His assimilative merging with the cat invites a discussion on posthumanism and/or transhumanism that I am unable to attend to in this chapter, however, what is salient is to what extent the narrator experiments with non-normative catlike (feral and bestial) subjectivity:

> what I really liked was walking in the town at night. Or between towns – industrial estates and bypass roads and canal routes. With Sara [the cat]. She made me see things I couldn't see on my own. Telegraph wires like webs stretched out between buildings. Splinters of glass across empty windows. Pieces of electricity generators left out to blacken. Things moving on the ground until the rain nailed them. The silver. The red. I used to stop in doorways, fall asleep, wake up with an erection and a mouthful of dust. (Lane 2006, p. 32)

This is not a man who is enjoying a normative subjecthood, rather he is exploring a queered hybrid feline identity within the detritus of the post-industrial urbanity of Oldbury. This exploration arouses him, and compels him towards his first queer (as in, same-sex partnering) the overlooked spaces, and the way its filth 'blackens'. This relishing clearly has a sexual element too, his exploration of his space and his self, in catlike ways, is a queer exploration. His first queer sexual experience. He undertakes this event as a feline would, and almost as an out of body experience where 'I seemed to be watching myself through a mirror: crouched over him, my fingernails dug into his sides, my head down like a cat lapping up milk' (Lane 2006, p. 33). And as his encounters increase, he takes on more of a feral selfhood that is bound to the filth and detritus of the post-industrial landscape. He contracts crabs and scabies from copulating in filthy alleyways and canal towpaths—a very dark and dispossessed view of the Urban Gothic. The Narrator's queer (in all ways) emergence is nurtured and specific to both his bond with Sara, the cat, and with the post-industrial, Black Country environment. This behaves as a very different Metropolitan LGBTQi space and culture, one where queerness necessarily flourishes in the uncanny and on the threshold. The reader follows as the narrator becomes a hybrid cat–human and as such his individuality of experience and his sexual behaviours cannot be read as normative. Anya Heise-von der Lippe's (2017) *PostHuman Gothic* has most recently addressed the posthuman hybridities in the Gothic. She argues that proliferation of narratives with cyborgs and inter-species creatures suggest a posthuman preoccupation in new Gothic texts that 'challenge the basic paradigms we associate with being human', (Heise-von der Lippe 2017, p. 1) and Lane's 'Scratch' does just this by having his narrator assimilate with a cat. In a literary return to Kipling's Just So Stories ('I am the cat that walks by himself, and all places are alike to me')—Lane closes the tale with his narrator confirming his hybridisation: 'I can seek out food and keep myself clean. I like people, but I don't need them. And I'll never trust anyone. My claws are sheathed. They're deep inside me' (Lane 2006, p. 39).

'My Stone Desire' and 'Still Water'

Lane takes the idea of assimilatory behaviour further in his short story 'My Stone Desire' from his *Where Furnaces Burn* (2012) collection. A central premise of this chapter is that a person outside of canonical metropolitan LGBTQi queer cultures—in this case, Wolverhampton—does in fact, experience their sexual identity in very different terms. And this story goes a long way towards arguing that the new Urban Gothic offers fruitful ground for these non-normative sexualities to develop. If 'Scratch' narrated a story of posthuman sexual hybridity with a cat, 'My Stone Desire' narrates a posthuman story of sexual hybridity with an environment. The story details how the landscape gives rise to a non-normative sexual identity that borrows from spaces where green and grey meet and where the Urban Brownfield spaces wrestle with decay.

Typical of Lane, the topos is described as a borderless space with limitless potential for liminal experiences: 'just where the factories gave way to fields and woodland, there was a low railway bridge of blackened stone and crisscrossed iron girders.... Young couples went there to smoke dope, drink bottled beer and screw' (Lane 2012, p. 4). Similarly, to be queer is to be part of that liminal realm. To be queer does not automatically make you a part of Metropolitan LGBTQi communities—it is not automatically assimilationist. Indeed, to be queer is to seek a political space that is removed from such discourse. Following Calvin Thomas' Lacanian readings of queer culture, if what is being sought is the 'sinister, yolky, excessive, excremental, obscene overflow of the "pure real"' that was forfeited in the cutting of the umbilical cord to be us with our 'unfindable and unspeakable anatomical complement' (Thomas 2008, p. 180), then, surely, such sexual subjects are sexually indiscriminate in their pure libido state. Queer communities seem to have this knowledge more than most communities and thus will happily break all the normative codes and cross gender binaries to make metamorphosis to the 'pure real' possible for all, acting as an attack on the heteronormative authority of intrinsic identity being based on gender binaries. And these queer bodies can appear to the normative as perverse in their feral, unruly, polyamorous and polymorphous behaviour.

In 'My Stone Desire' the topos has more to do than previous uses of Black Country landscapes in Lane's stories do. The town is a place where human copulation and reproduction are driven and assisted towards orgasm by the mechanised surrounds: 'I felt the distant approaching of a train. Then its passing shuddered through us, and the quiet was torn apart like a tarpaulin over a nail bomb. Kath pressed against me, breathing hard. My fingers found her open … it felt unreal, or perhaps more real than I was' (Lane 2012, pp. 4–5). Such an encounter suggests non-normativity, in so far as it is beyond human—it is an orgasmic fusion of the train, the man and the landscape. Later in the story, the protagonist revisits the site under the bridge and observes: 'I saw the bridge had already deteriorated: a dense black mould was spreading on the walls and blurring the overhead girders. There was a smell of decaying stone, if stone could decay' (Lane 2012, p. 6). There is more going on in this place than just nature invading the dank recesses of an abandoned post-industrial town—Lane's narrator reports on actual stone decaying. This Victorian industrial masterpiece—built to last millennia—is breaking down, disintegrating, sullied and (as the stone forms the structure of a railway bridge) unsafe. 'It's like', reports the protagonist, 'there's something dead in the wall' (Lane 2012, p. 6). An observation that does more than pull the situation into the realms of the Gothic, but qualifies it as peculiar; as queer. Like queer subjecthood that Lee Edelman argues, 'refuses the calcification of form that is reproductive futurism' (Edelman 2004, p. 48), the dying wall risks leaving no legacy—it is, after all, an object not an inhabited subject despite how much Lane anthropomorphises it. Similarly, and as Paulina Palmer has observed, traditionally Othered queer characters are 'not yet subjects but who form a constituency outside the domain of the subject' (Palmer 2007, p. 52)—in this sense they are threshold characters on the limen of becoming. Lane's characters are too in a process reproducing and becoming, within the queer and strangely alive-but-dying post-industrial urbanity of a decaying Black Country town. Such a site seems to sit outside the natural impulses of the human–animal, yet links their erotica with rot, decay and a kind of super-nature. 'My Stone Desire' describes a sexuality that is alternate and anterior to nature.

This space of strange sex is latterly revealed to be constructed from the assimilated and fossilised remains of various sexual encounters—'the structure of the bridge was made up of tightly packed, naked human bodies, twisted together in the warmth of slow decay. They looked as if they were about to move, but they were still. I was close enough to smell them' (Lane 2012, pp. 4–7)—an entombed site of both necrosis and orgasm. The bridge is, therefore, drawn in the mode of the Gothic sublime—a horror that characters are both drawn to and revolted by; a supernatural site where the locality comes to feed their addictions.

The story, 'Still Water' also from Lane's *Where Furnaces Burn* (2012) collection, provides a similar narrative of emergence set in Wolverhampton where decay and development meet and give rise to quasi-eroticism: 'The houses were built on either side of a railway bridge that had been condemned in the fifties, but never demolished. They backed onto a patch of wasteland where old canals had leaked into the soil, giving the landscape a fertile variety of plant growth and a pervasive smell of stagnant water. It made me think of unwashed skin' (Lane 2012, pp. 9–10). As with his other stories, sex, lust and intimacy are foregrounded against an uncanny background. Sex is again queered—here it is both out of doors and sadomasochistic—in the sense that 'queer' is the umbrella term for non-normative sexual mores and encounters. The sex act is a violent one that leaves each participant marked and struggling to breathe, yet is also tender and concludes with the lovers holding

> each other more tenderly than we had all week. At that moment, I recognised the cold fever-smell of stagnant water … I could just see a number of pale tubes hanging down from the water surface.… They were connected to long, translucent maggots that jerked in the water. My finger touched

> one of them. I threw myself backwards and stood there, breathing hard and trying not to vomit. (Lane 2012, p. 13)

Such a close affiliation between copulation and necrosis seems horrifically out of place, but also out of time. We are reminded that queer or queered time exists outside of the heteronormative experience and occurs in alterity.

Halberstam uses Kristevan terms to suggests that notions of time and space are essentially formulated through a heteronormative prerogative, arguing that, 'Reproductive time and family time are, above all, heteronormative time/space constructs' (Halberstam 2005, p. 10). This, they argue, results in codes of time and space becoming organised around the requirements and desires of heteronormative family life—where reproduction leading to child-rearing underwrites what it is to be a family. Many queer bodies either eschew or are forced outside of these time/space codes as a result of prejudice and otherment. The Queer Community has, as Halberstam observes, constructed 'new life narratives and alternative relations to time and space'; they have re-made 'community in relation to risk, disease, infection and death' and this has us all rethink 'the conventional emphasis we place on longevity and futurity' (2005, p. 2). '[Q]ueer subcultures produce alternative temporalities by allowing their participants to believe that their futures can be imagined according to logics that lie outside of those paradigmatic markers of life experience'—queer time-space occurs when alterity produces needs that exist anterior to normative time-space. 'Queer' in this formulation extends beyond the subjectivity of LGBTQi cultures, to umbrella all non-homonormative, outsider culture that, like queer bodies and queer communities 'make room for themselves in a crowded world' (Halberstam 2005, p. 20) by cultivating their difference.

'Wake up in Moloch', the Last Gothic Horror

There are many recent queer theorists who criticise the normative urban safe space, arguing that some queer subjects have carved out satisfying space for themselves in more parochial settings. However, Johnston and Longhurst argue that living in (which privileges a choice) and/or carving out (which indicates a need) non-normative or queered time-spaces wherever you may locate yourself and build a community, is paradoxically both a freedom and a constraint. Such spaces, she describes in tropes of the

Gothic sublime and as spaces of simultaneous 'dread and delight' (2010, p. 80). This is particularly pertinent to Lane's stories as presented here: non-normative locations that sit on queer threshold and are potentially monstrous and yet emancipatory potential. Specifically, and in 'Wake up in Moloch' what is encountered in these non-normative locations is Queer Hauntology—a space-time that is both within and without time; concrete and ethereal.

O'Rourke reminds us that humanity's primal urge to survive and continue, is a legacy of Promethean idyll and one that, since the Enlightenment, has been conceived of as a superior 'right' of the human: '[o]ccupying time and space is a form of persisting with others, of asserting a right to (corporeal) persistence, presenting an obdurate challenge to chrono-bionormativity which would seek to dispossess those very bodies' (2014). Poshumanists eschew this as arrogance and as the heresy of humanism. And Lane's 'Wake up in Moloch', from *Where Furnaces Burn* (2012) wrestles with this argument within the contexts of a very queer kind of haunting. Set in Dudley, an urban centre that grew out of the Industrial Revolution, Lane's narrator recounts how the town 'owes much of its life to the machine'(Lane 2012, p. 183) and the dark, satanic mills that put it on the map. '[I]t was inevitable', he foreshadows, 'that sooner or later, we'd have to give something back' (Lane 2012, p. 183). An account of a series of local murders unfolds, all linked to local gangsters and all cabalistic in nature. The reader is lead to an old factory in the outskirts of Dudley. It contains a strange and old machine. Watching the machine being worked by the many custodians that tend to it, the protagonist becomes aware that it is 'alive' and struggling to continue. Man is no longer alone of all the species in his desire to persist—he is not fully in control. Indeed, he has become enslaved by the very industry that his ancestors built in the Victorian area; the very machines that have since destroyed our planet and who's residue and legacy now threaten to destroy our species. *This* machine has evolved and it haunts in very queer way: it wants something back, something sexual and ceremonial:

> some kind of machine occupied most of the floor. It looked rather like a giant steam engine turned inside-out, with pistons and wheels and chains moving at the surface. Its black chassis was ribbed and pitted like a fossil. There were dozens of workers around it, holding onto its various surfaces and protrusions. They were all naked … the machine stroked them, oiled

> them, tore them. Their blood ran down their bodies and the sides of the machine, onto the dull floor. (Lane 2012, p. 189)

The ceremonial and the erotic combine in a depiction of the machine's vampiric fulfilment. The machine that put Dudley on the map has returned, haunting us with its desire to persist to, perhaps, assimilate via a very queer, sadomasochistic sexual copulation and thereby sustain itself. This is more than a queer haunting and a ghost collecting a debt, this is a narrative that questions what it is to be human and challenges the predominance of the human—one of the central anxieties of our new millennium.

The canal paths of 'Scratch', the non-places in 'The Lost District' the disused factories of 'Wake up in Moloch' are where Lane's protagonists embrace the anxiety-ridden threshold spaces of the de-industrialised Black Country. The Black Country, itself, acts as a queer subject. Lane's stories are preoccupied with a region and subjectivities in positions of flux, contradiction and transition—with marginal people and places that are struggling to come to terms with post-industrial upheaval. His urban spaces are, by turns, horror-filled and Gothic. They represent the scenography the Anthropocene at the millennium and our anthropic legacy in their post-industrialised imagery: dilapidated housing estates, decaying factories, abject enterprise zones and retail sites. In such queer thresholds, Lane draws out subjects that are gender-fluid, shape-shifting, assimilative and eroticised.

References

Aguirre, Manuel. 2017. Thick Description and the Poetics of the Liminal in Gothic Tales. *Orbis Litterarum* 72 (4): 294–317.

Campbell, Ramsey. 2016. Afterword: The Whole of Joel. In *Something Remains*. Staffordshire: The Alchemy Press. Kindle ebook.

Coleborn, Peter. 2016. Foreword. In *Something Remains*. Staffordshire: The Alchemy Press. Kindle ebook.

Edelman, Lee. 2004. *No Future: Queer Theory and the Death Drive*. Durham: Duke University Press.

Halberstam, Judith. 2005. *In a Queer Time and Place: Transgender Bodies*. Subcultural Lives, London: New York University Press.

Haughton, Hugh. 2003. Introduction. In *The Uncanny*. London: Penguin Books.

Heise-von der Lippe, Anya. 2017. *Posthuman Gothic*. Cardiff: University of Wales Press.

Johnston, L., and R. Longhurst. 2010. *Space, Place, and Sex: Geographies of Sexualities*. Maryland: Rowman & Littlefield.

Lane, Joel. 2006. *The Lost District and Other Stories*. San Fransisco: Nightshade Books.

———. 2012. *Where Furnaces Burn*. Hornsea: Drugstore Indian Press.

Morgan, Chris. 2016. 'Joel Lane, Poet', from *Something Remains*. Staffordshire: The Alchemy Press, Kindle ebook.

O'Rourke, Michael. 2014. Time Tangles. In *Time Binds*, from Social Text Online. https://socialtextjournal.org/periscope_article/times-tangles/. Accessed 05 August 2017.

Palmer, Paulina. 2007. Queer Transformations: Re-negotiating the Abject in Contemporary Anglo-American Lesbian Fiction. In *The Abject of Desire: The Aestheticization of the Unaesthetic in Contemporary Literature and Culture*, ed. Konstanze Kutzbach and Monika Mueller, 49–68. Amsterdam: Editions Rodopi.

Stein, A. 1992. Sisters and Queers: The Decentering of Lesbian Feminism. *Socialist Review* 22 (1): 33–55.

Thomas, Calvin. 2008. *Masculinity, Psychoanalysis, Straight Queer Theory: Essays on Abjection in Literature, Mass Culture and Film*. Basingstoke: Palgrave MacMillan.

Abjection and Anime in the Anthropocene: Amano and Oshii's *Angel's Egg*

Kwasu David Tembo

Conceptually, the Gothic has had a history of being a space of liminality and play and has always facilitated encounters with otherness, both in the world and in ourselves. In considering the New Urban Gothic on the edge of the millennium, this essay will explore how Mamoru Oshii and Yoshitaka Amano's 1985 opus *Angel's Egg* (天使のたまご *Tenshi no Tamago*) represents and reimagines critical Gothic studies, providing a vivid, reflective aesthetic in which to consider what it is to be human. While traditional Gothic spaces—cemeteries, moorlands, abbeys, ruins, castles, and manors—facilitate societal and individual self-reflection through encounters with super-nature embodied as 'Other', Oshii and Amano's anime illustrates that this same space can facilitate contact with self-reflection without the intercession of monsters and/or revenants of any kind. *Angel's Egg*'s aesthetic and narrative qualities are situated within Japanese anime, Oshii's own corpus of directorial work, and the medium's engagement with Gothic tropes and aesthetic. These will be explored here, via close readings of the visual and symbolic Gothic elements of Oshii and Amano's text. This interpretation draws on Julia Kristeva's

K. D. Tembo (✉)
Independent Researcher, Harare, Zimbabwe

H.-G. Millette and R. Heholt (eds.),
The New Urban Gothic, Palgrave Gothic,
https://doi.org/10.1007/978-3-030-43777-0_5

discussion of abjection to examine how Oshii directs his anime in combination with the Gothic and the surreal to represent the Gothic urban space as one of existential stasis, crisis, and reflection.

Angel's Egg is a rare collaboration between two titans of Japanese audiovisual art and is notable for its slow and mournful pace, its surreal Gothic aesthetic, and its sparse dialogue. The plot follows an unnamed Girl who lives alone in a nondescript building abutting an abandoned techno city painted in Gothic hues of blacks, blues, and greys. She spends her time minding a large egg which she protects while scavenging for food, water, and glass flasks. However, the film does not open on her, but with a Boy, in military attire, carrying a cross-shaped weapon watching a giant orb-like vessel descend from the sky. This, we discover, is a contained cathedral city covered in Goddess sculptures and fossilized remains. The descent of the orb wakes the Girl who begins her routine scavenging. She encounters the Boy who has entered the city in a biomechanical tank—part military machine and part arachnid, this tank is notable in that it is drawn in an angry red-orange colour that glares against Gothic greys and purples. She is frightened off by him and so relocates to continue her scavenging alone, avoiding contact with any anthropic occupants of the city (the Boy and harpoon-wielding fishermen-soldiers). When the two meet again, the Boy produces the egg the Girl had been protecting but had left momentarily. He cautions her to internalize the things she holds dear lest she lose them and then asks what she believes is inside the egg. The Girl answers that she cannot tell him. The Boy suggests breaking it to find out. This angers the Girl and drives her away again. After pursuing her, the Boy eventually wins her trust by helping her evade a ritual 'hunt' for coelacanths carried out by the fishermen-soldiers. These chase gigantic shadows of coelacanths that swim across the surface of the streets and the edifices of the buildings in a scene reminiscent of Margritte's art (Image 1).

Oshii and Amano's description of the primary characters as 'the Boy' and 'the Girl' take on an added significance when considering the text as a meditation on abjection. Being that the text's primary characters are masculine and feminine, their relationship and the latent dynamics of power, patriarchy, and privilege surrounding their respective interest in the text's central object (the egg) suggests a few things from an EcoGothic and Post-human perspective. From these critical positions, Oshii and Amano move beyond the separate gendered spheres division

Image 1 Coelacanth shades swim across the city

of power and fecundity of emotion to show both characters undifferentiated in the mystery of the egg. Both are shown to be psycho-emotionally damaged by the abjection they feel as a direct result of the egg—a point of interest this essay will return to later to discuss in depth. At the end of the first third of the narrative, the Girl tells the Boy she intends to hatch her egg. As the pair rest in the Girl's hovel, the Girl falls asleep. While unguarded, the Boy takes the egg and smashes it with his cross-like weapon and leaves. On waking, the Girl is horrified. She pursues the Boy and, in her haste, falls into a ravine and into a surreal subterranean water-realm where she is transformed. Her final breath rises to the surface, releasing a multitude of eggs.

The film has, in turns, been described as: 'a lyric film' (Ruh 2004, pp. 46–47); 'a surreal and Zen film' (McCarthy 2009, p. 39); 'a haunting poem of melancholic science-fantasy' (Willis 1997, p. 20); and 'an animated tone poem featuring ... dimly lit urban tableaux filled with dilapidated architecture, shadow figures, and hollowed-out symbolic forms' (Suchenski 2004, n.p.). Suchenski goes further, describing this as 'Oshii's most personal film' (Suchenski 2004, n.p.) and, in interview (Horn 1996) Oshii admits this, and the importance the Judeo-Christian narratives (Bible Stories of his boyhood) have had on his aesthetic philosophy, and the iconography and *mise en scène* of *Angel's Egg*. The film is very much in keeping with Oshii's predilection for complex ontological, existential, and psychological stories and social and cultural debates that typically conclude with neither an affirmation of hope nor a promise of deep despair (Cavallaro 2006). As such, his film 'challenge[s] the placid formulaic assumptions of conventional animation and because they constitute a unique, deeply personal, and increasingly relevant meditation on the changing face of personal and national identity in the 20th and 21st centuries' … offering the reader 'a sustained meditation nature of personal identity, the ways in which the frontiers of being are transformed by active searching and constant change' (Suchenski 2004, n.p.). Typical of anime, the film remains open to a variety of readings, however, this chapter will explore the New Urban Gothic reading of its text as a vivid existential desperation and collapse of faith at the end of humanity. While the relationship between the loss of faith, the Gothic, the urban, and the surreal will be addressed, it is acknowledged that the symbolic and thematic overlaps between *Angel's Egg* and the Judeo-Christian faith are too numerous to detail in full here and are well considered elsewhere. This essay begins with a critical discussion of *Angel's Egg* with a reading of the display form that anime is known for because the Gothic scenography and the affect of the Anthropocene in *Angel's Egg's* images are striking and crucial to the Gothic nature of its city-space.

The Gothic Anime City: Space and Affect in *Angel's Egg*

Angel's Egg is mostly without dialogue and so surreal in its narratology. The clearest way to gain an impression or sense of the text *as* a film is via its powerful visuals. The film's aesthetics depict a world in decline. Inundated by both water and shadow. The film relies on chiaroscuro (the artistic convention of strong contrasts created by dark and light opposed) and saturnine *mise en scene*, but one that is unfamiliar to the everyday of the viewer. The aesthetic rendering of the city in the film represents it as a 'character' unto itself. The city is empty but for its scenography: winding alleyways, checkerboard deserts, abandoned antennae reaching upward like techno forests, eerie lampposts, and foreboding edifices, which evoke the sensation of time interrupted. Oshii achieves this effect through the use of static shots, and lingering takes developed with extremely slow camera tracking and panning. The result of which is the feeling that neither diegetic (story's world) or extradiegetic (viewer's world) time has, or ever can, elapse. Every frame is flooded with a sense of oppressive stillness and a disquieting suspicion that the City itself is, panoptically, watching the protagonist(s).

A brooding paranoia pervades and is exploited by Oshii's framing techniques. Portals of both observation and sequestration are everywhere: windows, service tunnels, giant industrial exhaust vents, and archways that stare with black or clouded voids for eyes. Perspective is manipulated to sustain a sense of both danger and observation relative to the object of surveillance. This is a typical strategy to empower the displayed image, and here it used to eschew any sense of a homely invitation, replacing it instead with the *unheimlich* feeling of the uncanny (Freud 1997) and the edgy feeling of being watched (Image 2).

This visual example of the uncanny is a quintessentially Gothic representation, just as the ruin of the urban is a quintessential representation of the ruin we have made of our lived era—the Anthropocene. Dilapidated, semi-ruined edifices, buttressed with Gothic spires, iron gates, moribund fountains, and an encroaching vine-like filigree, present a city

Image 2 Windows invite but also threaten

in unfinished decline. And hybrid revenants are sewn into the very architecture of the crumbling city where ghostly gargoyles suddenly reanimate as militarized fisherman when hunting the spectres of shadow-fish. Both the hunter and the hunted are surreal ideas of abandonment—abandoned or lost culture and industry. Only the Girl and Boy seem to be alive in this otherwise dead city. Everything from its mechanical infrastructures and cultural centres such as cathedrals, town squares, and even homes are all abandoned. The fishermen and their fish prey only amount to a shadow-play within the surreal. There are numerous ways of reading this, one interpretation is that this is Oshii and Amano's commentary on both the intractability and pointlessness of desire. The fisherman perform this when their harpoons strike only shadow when hurled at the shadow-fish gliding on or beneath the surface of the city's silent edifices, and is supported later in the narrative, when the Girl discovers that the egg she so rigorously incubates and protects is hollow. In each instance Oshii and Amano present visual metaphors for the idea of the meaninglessness and/or disappointment with the pursuit of longing and desire.

With its stasis, non-activity, lack of people, and generally purgatorial silence, there is an implication here of a city being, in some way, temporally immortal. This is a conversation that its surreal design engages in, and in such a way as to draw the space as both heterotopic and frozen, but also anticipatory, such that time and space do not function in the way we typically understand them to. The term heterotopia comes from Foucault, in which the author theorizes on various phenomena in a conceptual merger between the heterotopic and the absurd. Here, it refers to the idea that heterotopic space or heterotopias are concepts in human geography elaborated by Foucault and then other post-structrualists to describe places and spaces that function as non-hegemonic or spaces of otherness (Foucault 1967). They are simultaneously physical and mental, such as the space of a phone call or the moment when you see yourself in the mirror. In this way, heterotopia refers to spaces typified by sociopolitical and cultural praxes. All flows of power and ideology of a given culture, break down or are renegotiated within them. The term describes the human geographical phenomena of spaces and places that function in non-hegemonic ways or conditions: they are spaces of otherness, liminality, fusion, confusion, play, and dynamism. The heterotopic in *Angel's Egg* is a hybrid of various modes and reflections of matter: physical, ephemeral, paradoxical. In the film, crisis, complex ontological conditions, searches for meaning, and the feelings of nihilism, stagnation, ennui, all pervade in an Urban Gothic landscape. The Girl's home, for example, is as a museum/mausoleum or graveyard where bones and brick knit together. This combines antipodal concepts—the ancient and the modern; the living and the dead; the playful and the Gothic—that are then ossified and meditated within the fabric of the city (Image 3).

Perhaps the single-most notable animation technique in the film is its use of negative space. Its representation and play with negative space is often in relation/antagonism to the Girl, whose bright white hair and bright white egg both cut out sharp emptiness in the umbras—the fully shaded inner region of a shadow cast by opaque objects—of the Gothic *mise en scène*. Further, Oshii almost entirely eschews gradient in favour of stark chromatic delineations: be it black and red as it is on the checkerboard desert; red and blue as it is with the gargantuan eye and the sky in which it sits; or any number of the blue, violet, grey, green, and/or black settings in which the characters find themselves. While shadow and stark colour are used to evoke visual and emotional feelings of separation, sequestration, isolation, exile, paranoia, and surveillance, there is one constant symbolic element that *binds* these aspects all together—water. Oshii's use of water, in various states of motion (running, pouring,

Image 3 A city of bones and ossification

bubbling, raining, pooling, etc.) pervades every aesthetic element heretofore described. Not only is water a symbolic frame for the city, as the end of the film reveals, but it is manifest in most, if not all, objects within it. Two of many examples include the Girl's billowing hair, and the twisting tributaries making up the Boy's cross. Water also serves a foreshadowing function. Early in the film during a scene where the Girl drinks from a river near the edge of the city and fills yet another water jar, the surface of the water, with its mirror-like blackness, speaks of self-reflection, self-understanding, and ignorance of purpose or self. However, there are no clues as to who she is, why she is in this city, the nature of the city, or, perhaps most mysteriously of all, why she is bound to both the city and the egg she nurtures. Furthermore, the black water of the river reflects no image, no inkling of the emergence of something beyond the city or of the city itself. Yet the city rises like the reflection of the girl herself, in the film's penultimate moments. Framed from below the surface or above it, the attributes of the water remain unchanged and, like the binding nature of water, the Girl's fate is both bound to the element and led by it. She, collecting water in the city, is surrounded by it, threatened by it, inundated with it, reflected in it, and fears her own destruction by it. In this way, the Girl, like the city is captivated or laid under siege by the water's ability to reflect and destroy. At the end of the film, the water indeed rises to engulf the city, much like the Biblical story of Noah and the Ark or the Greco-Roman story of Deucalion and Pyrrha. A deluge that carries with it connotations of destruction renewal, cleansing, purification, and (re)genesis.

The Gothic City and Anime: *Angel's Egg*'s Relationship to the Gothic

Angel's Egg may bear an obvious relationship to ecocritical discussions of cultural products and the Anthropocene, in its blatant depiction of human ecological destruction, however, this discussion of the Urban Gothic and, in particular what is new about it in this film, is centred around space and, specifically, a non-place topos. *Angel's Egg* is an interior place mapped by a surreal topos. Rather than rely on a latent assumption that Gothic aesthetics and narratological tropes, traits, and leitmotifs are static, recursive, and/or universal, I have argued that the myriad ways gothic tropes have been deployed in various media traditions reterritorializes Gothic themes and concepts in this anime. The concept of topophobia is helpful here in theorizing the confluence of internal and external space, such that the fears of the former manifests, albeit abstractly and in a surreal way, in the Gothic space of the latter. Jane Chi Hyun Park (2011) offers some helpful insights into the Gothic within an Eastern audiovisual tradition that service this argument. For Park, the quintessential conventions of the Gothic genre are employed in contemporary Asian films and literature in ways that go beyond a melding of gothic tropes with realist themes, aesthetics, and concerns. Park reminds us that the Gothic 'is not exclusive to the West but is also deeply embedded in Asian cultures, where fantastic 'others' serve as metaphors for different kinds of [cultural] fears and anxieties' (Chi Hyun Park 2011, p. 107). The Gothic expresses itself along numerous other non-textual meridians in the East, such as fashion. For example, and as Tom Looser (2009) observed, Gothic style (especially in its connection to a 'Lolita' fashion) in Japan had reached such a level of mainstream culture at the millennium as to spawn widely varying and dissimilar subgenres (Gothlolita, Vampire Lolita, Steam-Punk Lolita, Black Lolita, Victorian Lolita, etc.) that have only vague associations with the qualities typically attached to the European Gothic era, but whose central characteristics are relevant to new Urban Gothic narratives (Looser 2009, p. 55). In *Angel's Egg* an anonymous *animated* space without recognizable totems of culture or history subverts the traditional Urban Gothic scenography as deployed in a new form—surrealistic Gothic Anime. The *Angel's Egg* topos is untethered to Gothic identity constructs and meanings and so inverts the audience's expectation of typical Gothicism through a surrealization of Gothic's most conventional tropes. *Angel's Egg* might, therefore, be accurately described as interpolating the Gothic with the intention of subverting its various aesthetic and narrative conventions. It centralizes but also surrealizes the affective experiences of the psychological and emotional space of the text and its protagonists, without resorting to signifiers such as the castle, the fog, or the monster (Marshall 2017). Yet, implicit in Oshii's Gothic appropriation, bricolage, and frame of reference is a focus on exploring both the 'disempowered and disjointed subjectivity' of the Girl (Mettifogo 2010, p. 74).

Another way in which Oshii destabilizes Gothic tropes pertains to movement. Donovan-Condron points out that 'while travel is a frequent trope of Gothic literature, it is usually more or less in one direction, away from home and toward danger' (Donovan-Condron 2013,

p. 690), *Angel's Egg* is constructed as a surreal urban Gothic dead-end. While, on the one hand, the film is a traditionally Gothic in that looks 'simultaneously back and ahead, both in literature and in history, the Urban Gothic demonstrates that the individual is not just a part of the whole, but apart within the whole' (Donovan-Condron 2013, p. 685), on the other hand, any sense of boundaries is moot as border-crossings, distinctions, and transgressions are not physically possible. Following Fred Botting's understanding that the implications of psychological and emotional transgressiveness incubated by the Gothic permeate topographical distinctions—

> Gothic terrors activate a sense of the unknown and project an uncontrollable and overwhelming power which threatens not only the loss of sanity, honour, property or social standing but the very order which supports and is regulated by the coherence of those terms [such that] the line between transgression and a restitution of acceptable limits remains a difficult one to discern. (Botting 1996, pp. 5, 7–8)

—Oshii and Amano push the terror in *Angel's Egg* by questioning humanity's tenets and intent *in principium*. What is communicated in *Angel's Egg* is not a terror and evil that exists in the world and/or in the everyday, but an empty terror and evil with no provenance, no purpose, and indistinct from any other human phenomena. This is the nihilism of modernity—the 'God is Dead'; the universe is in chaos principle—that leaves us with a pervasive feeling of emptiness and sometimes terror.

Following Sara Wasson's argument that a reterritorialization of Urban Gothic literature's predominant physical location within the context of the city simultaneously reterritorializes the psychological and emotional threats Gothic characters encounter, the landscape of the urban in *Angel's Egg* has no need of supernatural terror as 'the metropolis evokes the horrors of human violence and corruption [which is] terrifying enough' (2010, p. 30). The film suggests that the loss of the distinction between self and Other is both inevitable and traumatic as it results in an abjection. Here, the surrealist Urban Gothic—its highly abstract psychological and emotional and/or symbolist constructions, figurations, and motifs—behaves in such a way as to render the psychological and emotional space of the psyche and the solipsism of its protagonists, topophobic; a phenomenon of anxiety (Kelly Gray 2009). As solipsism holds that the mind exists because it can reason only of itself, the solipsist believes that anything outside or beyond one's own mind is unsure to exist. When considering *Angel's Egg*, solipsism is central to the protagonists' anxiety is the claustrophobic space in the film, steeped as it is in abandonment, apostasy, nihilism, and abjection. While the Girl's protection and minding of the egg seems to suggest a purpose, it is one riddled with uncertainty. The Girl admits to the Boy that she is unsure what or where the city is, what happened to the creatures that existed there before, how their remains were fossilized, or what the egg is, what is inside it or where it came from. In this sense, her routine of collecting water and guarding the egg would appear to be habitual and ritualized methods of subduing any anxiety she may have concerning the uncertainty and ambiguity pervading everything. This results in apostasy—an abandonment, loss, or renunciation of a central belief as a result of uncertainty and doubt, which cannot be ameliorated through routine and ritual. While typically used in reference to specific political and ecumenical praxes, here it is used more generally to describe the ennui that follows the protagonists' loss of purpose. This is the reterritorialization that occurs of which the protagonists cannot escape—the city in *Angel's Egg* is a site of unfinalized finality; a terminus without termination; and end without an end.

Robert Mighall argues that the notion of the Urban Gothic is more than simply a work set in a city, but a story in which a city performs as a repository or 'concentration of memories and historical associations that are found in architectural and topographical forms' (2007, p. 57). Implicit in this definition is a sense of historicity, inheritance, iniquity, and enclosure—arguably, creating an abject space. While Kristeva in *Powers of Horror* explores various aspects of abjection, drawing primarily from the psychoanalytic works of Freud and Lacan, central to her argument is the relationship between human beings and their reaction(s) to perceived threats of the dissolution of meaning. Kristeva's theorization of abjection is pertinent and helpful here because it helps explain the ways in which the psychological and emotional processes and consequences of a loss of faith can manifest not only in the individual experiencing said processes but symbolically in the objects and spaces in which these processes occur. In other words, Kristeva helps provide a psychoanalytic reading of the topos of *Angel's Egg*. In her text, Kristeva argues that the abject produces an effect of horror based on the uncanny disruption of the familiar. The first object of this horrific uncertainty is the individual sense of self, undermining the myth of its singularity and sovereignty, resulting in psychological and emotional dread and anxiety. Paradoxically, the abject produces not only a sense of awareness of something repugnant within one's self but a compulsion to repeat the experience. The abject and its uncanny action operate surreptitiously through both the aesthetics and narrative of *Angel's Egg*, most significantly through the active space of the Urban Gothic. This is especially true of the relationship the Girl has with her egg, which can be described as one of abjection: a thing is perceived as loathsome or horrifying, and in which the 'I' attempts to expel it from itself (Kristeva 1982, p. 3). The egg, a leitmotif of doubt, is simultaneously nurtured and expelled. Both the city and the egg, as seen against the Girl, mark the fragile division between the inside and outside of humanism and the human—producing a disruption of identity that Kristeva describes as 'a radical evil' (1982, p. 70). The egg then, is a symbolic amalgam of the desire associated with life and the prospect of death and waste—what Kristeva's describes as 'the other side of the border, the place where I am not and which permits me to be, the corpse, the most sickening of wastes' (1982, p. 3). The implication here is that the egg is less than a casket of stillbirth or death and—more radically, perhaps—an incubator of nothingness. The city, the egg, and the Girl form a triskelion of abjection sited that aimlessly repeats—reflections of which are the Girl's repetitive collection of water flasks, her scavenging, and her nurturing of the egg. Kristeva's thesis on abjection is centred on its association with the female body—and here the body of the protagonist is female, but not quite adult, she is a girl standing on the limen of womanhood. Thus, the city around her becomes abyssal, a marker of the 'genesis and the obliteration of the subject, for it is a space inhabited by the death drive' (Grosz 1989, p. 77). Conceptually, the surreal images of aggressive soldier-fishermen, the orb-like vessel, and the giant coelacanths further destabilize the notion of psychological and emotional borders and rules. The sense is that for the Girl nothing is possible, but for the surreal space, everything is possible.

This chimes with the familiar–unfamiliar *unheimliche* (uncanny) feelings of the abject evoked in the framing and the windows that as described earlier, and it is what Julian Wolfreys claims blurs the clear demarcation between the self and the other resulting in a sense of disorientation that threatens identity, bodily integrity, and the sense of the self (Wolfreys 2001, pp. 3, 239–245). Remembering that the uncanny is a mechanism of return, the compulsion to repeat, to relive a past experience; it is a return of the repressed, caused by an anxiety associated with the past (Freud 1997, pp. 193–195), the Boy's entry into the story seems just such an uncanny return. The Girl and the Boy also have an uncanny relationship with one another—they possess a familiarity with each other that is both unfamiliar and disorientingly frozen and the Boy's crushing of the egg calls to mind the 'anxieties about fragmentation, about disruption or destruction of personal stability, bodily integrity, and immortal individuality' that Bronfen (1992, p. 113) refers to as being the destabilizing factor of the Uncanny.

In lieu of narratological robustness, the aesthetic of the city as an Urban Gothic space *becomes* the narrative in *Angel's Egg*. It dominates the human landscape; so much so that it becomes 'the presiding entity, the ambit of fear' (Pine 1994, p. 169) 'a desirable and terrifying, fascinating and abject, maternal body' (Dadoun 1989, p. 54). Allyson Kreuiter's insights concerning the relationships and tensions between Urban Gothic spaces and the experiences of inhabitants in such spaces are helpful in elucidating the sense of ontological and existential states experienced by the Girl and the Boy. In the film, the surreal Gothic space of *Angel's Egg* is, fundamentally paradoxical: at once the locus of maternal nurturing and an abject space of morbid remains:

> the terrors of [the film's city's] past and present are dictated by this powerful urban Gothic locus, which usurps human consciousness and [free] will, and governs and controls the historical past, and the present strife that its inhabitants continue to re-enact. In [a] reformulation people...cannot be considered 'men and women any longer'. Rather, they have been 'unconsciously made part of the place, buried to the waist among the ruins of a single city, steeped in its values. (Kreuiter 2016, p. 73)

In this sense, the characters' inability to recall their identities, the history of the city, or their purpose within it is irrelevant as 'identity is merely a fiction, because everything is lived through the will of the city. Absorbed into the city, men and women become merely aspects of the city's Gothic otherness' (Kreuiter 2016, p. 73). If framed as a response to the contemporary human condition, the implication is that regardless of how new or Gothic or surreal an urban space is, it will always-already incubate problematic antisocial phenomena of ennui and nihilism. Further, and returning to the theme of surveillance touched on earlier, there is the unsettling feeling throughout *Angel's Egg* that any and everything that transpires is an elaborate experiment being conducted *by* the city and dictated, overseen, and overshadowed by yet another city—one that seems ambivalent towards the phenomena of anxiety it creates. At its heart, are questions of agency. Within the city's paradox is the undeniable sense that it has voided the Girl and the Boy of any agency, memory, or sense of self.

If Milton gave us Pandemonium, a capital city of Hell in *Paradise Lost*, then Oshii has given us 'Panabjectium', a capital city of Purgatory in *Angel's Egg*. The overwhelming sense of abject time—time frozen or suffocating—does not entirely nullify the feeling of dark life persevering in the city. This sense of stillness and the slow action of the film chimes with Homi Bhabha's description of time in Gothic texts—a 'bizarre temporality characterized by a stillness of time' (1997, p. 451). While revenant or recumbent, the city is still an alive, yet passively inert, surreal Gothic landscape suspended in time. Oshii and Amano have created a tale that refracts the uncanny strangeness of the typical. The Gothic supernatural does not appear as thematic or material in this space. Instead, the surreal Gothic city in *Angel's Egg* is a ruined living urbanity that lingers as a ghost (Beville 2013). As previously argued, Oshii and Amano do not simply surrealize the Gothic for its own sake, rather they invent a new means of literally and figuratively drawing a bridge between art, abjection, space, and the psyche. The text resonates with a dull ache of anticipation; *something* is always-already about to happen; a deluge is coming. There are also moments that evoke the violence of species extinction, such as the scene in which the fisherman take to the streets and attempt to harpoon shadows of long-gone fish. But, more pointedly, the Boy's destruction of the egg suggests the malevolent violence opposing the feminine topos—the Mother Earth—that will be our downfall. The city's death drive is noted in the disintegration of its space which, as a process, is expressed both as at once always-already complete, and always-already about to occur (Farnell 2011). In this, *Angel's Egg* warns of the fragility and precariousness of the age of the human and the abjection that ensues in the face of the loss of civilization, order, identity, stricture, and purpose. This new Urban Gothic is claustrophobic in its revenant always-already ending topos in crisis. Mettifogo asks 'does the Gothic limit its subversive powers to the staging of tensions and contradictions, followed by a return to normativity, or does it promote change and reformation?' (2010, p. 83). Perhaps, it does neither now? Perhaps, this is Oshii and Amano's comment: that there is no solution or resolution, only the concrete disorientation, affect and existential crisis that results. If *Angel Egg* creates anything, perhaps it creates a vivid representation of a new Urban Gothic that offers us a new aesthetic with which to express the abjection of the Anthropocene.

References

Beville, Maria. 2013. Zones of Uncanny Spectrality: The City in Postmodern Literature. *English Studies* 94 (5): 603–617.

Bhabha, Homi. 1997. The World and the Home. In *Dangerous Liaisons: Gender, Nation and Postcolonial Perspectives*, ed. A. McClintock, A. Mufti, and E. Shohat, 445–455. Minneapolis: University of Minnesota Press.

Botting, Fred. 1996. *Gothic*. London: Routledge.

Bronfen, Elizabeth. 1992. *Over Her Dead Body: Death, Femininity, and the Aesthetic*. Manchester: Manchester University Press.

Cavallaro, Dani. 2006. *The Cinema of Mamoru Oshii: Fantasy, Technology and Politics*. New York: McFarland & Company.

Chi Hyun Park, Jane. 2011. Ghostliness in Asian Diasporic Film and Literature. *Asian Studies Review* 35: 105–113.

Dadoun, Roger. 1989. Fetishism in the Horror Film. In *Fantasy and the Cinema*, ed. James Donald, 39–61. London: British Film Institute.

Donovan-Condron, Kellie. 2013. Urban Gothic in Charlotte Dacre's Zofloya. *European Romantic Review* 24 (6): 683–697.

Farnell, Gary. 2011. Gothic's Death Drive. *Literature Compass* 8 (9): 592–608.

Foucault, Michel. 1967. *Of Other Spaces: Utopias and Heterotopias*, trans. Jay Miskowiec. Paris: Architecture/Movement/Continue.

Freud, Sigmund. 1997 [1919]. The Uncanny. In *Sigmund Freud Writings on Literature and Art*, trans. James Strachey, 193–233. Stanford: Stanford University Press.

Grosz, Elizabeth. 1989. *Sexual Subversions*. Sydney: Allen & Unwin.

Kelly Gray, Elizabeth. 2009. The World by Gaslight: Urban-Gothic Literature and Moral Reform in New York City, 1845–1860. *American Nineteenth Century History* 10 (2): 137–161.

Kreuiter, Allyson. 2016. The Urban Gothic City in Lawrence Durrell's The Alexandria Quartet. *English Academy Review* 33 (2): 68–80.

Kristeva, Julia. 1982. *Powers of Horror: An Essay on Abjection*, trans. Leon S. Roudiez. New York: Columbia University Press.

Looser, Tom. 2009. Gothic Politics: Oshii, War, and Life Without Death. *Mechademia* 4: 55–73.

Mamoru Oshii interviewed by Carl Gustav Horn (1996). Listfiction, 11 February 2016. http://listification.blogspot.com/2016/02/mamoru-oshii-interviewed-by-carl-gustav.html.

Marshall, Bridget M. 2017. "There Is a Secret Down Here, in This Nightmare Fog": Urban-Industrial Gothic in Nineteenth-Century American Periodicals. *Women's Studies* 46 (8): 767–784.
McCarthy, Helen. 2009. *500 Essential Anime Movies: The Ultimate Guide*. New York: Harper Design.
Mettifogo, Mariarosa. 2010. The Unspeakable Violence of Gender: A Gothic Reading of Neera's Crevalcore. *Italianist* 30 (1): 69–86.
Mighall, Robert. 2007. Gothic Cities. In *The Routledge Companion to Gothic*, ed. Catherine Spooner and Emma McEvoy, 54–63. New York: Routledge.
Pine, Richard. 1994. *Lawrence Durrell: The Mindscape*. New York: St. Martin's Press.
Ruh, Brian. 2004. *Stray Dog of Anime: The Films of Mamoru Oshii*. New York: Palgrave Macmillan.
Suchenski, Richard. 2004. Mamoru Oshii. *Senses of Cinema*. http://sensesofcinema.com/2004/great-directors/oshii/. Accessed 2 May 2018.
Wasson, Sara. 2010. *Urban Gothic of the Second World War: Dark London*. London: Palgrave Macmillan.
Willis, Donald C. 1997. *Horror and Science Fiction Films IV*. New York: Scarecrow Press.
Wolfreys, Julian. 2001. *Critical Keywords in Literary and Cultural Theory*. Basingstoke: Palgrave Macmillan.

Urban Gothic: Ruin and Residue in the Anthropocene

Holly-Gale Millette

Since the turn of the last century, World's End narratives and/or Human's End narratives have proliferated. These have found their aesthetic politic in narratives of decadence, difference and degeneration. As early as 1931, Stuart Chase described the potentiality for urban cities to become negated landscapes. He called these landscapes 'Megalopoli'—places whose towering architecture and oases of cloisters of luxury provide a privileged buffer from the dust, noise, filth, congestion and inflated property-values of an unworkable financial system. We know that the resurgence of such representations occurs in periods of ontological and epistemological crisis, and that these most often foreground the oppositions between the Self and the Other—discourses in which that Other is menacing, evil or the Darwinian survivor in a 'ruined' world. As Wolfreys has reminded us, in his previous retrospective in this volume, this immediately locates the Gothic as multiple and as always in the process of othering itself—The Gothic, says Wolfreys, 'transforms and is transformative'. The essays in this section takes the Gothic into the millennium, insisting that the subject-focused heritage of the Gothic is now giving way to a Gothic that has transformed our earth and our species. This threatens precisely because of man's capacity to monstrously transform his environment.

In *Capitalist Realism*, Mark Fisher argued that the limiting horizons set by neoliberal society mean it is now easier to imagine the end of the world than the end of capitalism. In his second book, his last published before his suicide, *Ghosts of My Life: Writings on Depression, Hauntology, and Lost Futures*, Fisher argues that cultural time has stalled and we've become increasingly incapable of producing the 'new', the 'now' and postulating the 'next'. At the end of such a history, all that's left is an endless return of dead forms. In our millennial Gothic tales, what haunts, in part, is this endless return to dead forms (Broaks 2014). These are most often understood as uncanny forms in the Gothic, however Fisher discusses these forms as decentering because they are external arbitrary forces—not the internal ones Freud refers to. Fisher's posthumously published, *The Weird and the Eerie* (2017), explores those two titular concepts of the Gothic as radicalised narrative modes in the millennium that re-define concepts or moments of, what Bonneuil and Fressoz (2016) have identified as, the transcendental shock of the Anthropocene. Gothic narratives now move beyond the modernist and enlightenment philosophies on which they were built to, as China Miéville, when interviewed by Stephen Shapiro, put it, 'surrender' (2008, p. 67) to the arbitrary forces that shape the human encounter. Millennial spectres are no longer as benign as *Frankenstein* (1818), *The Mummy* (1827) or *Count Dracula* (1897). Acknowledgement that death and decay are all around us is not new. Freud isolated the compulsion towards self-destruction (Death Drive) as early as 1920, and others have since explored extensions of this in thano-tourism, poverty tourism, and undead role-play. What is new, is how self-aware the reader is of the systemic nature of the dark, the eerie and the weird—and how connected it is to Anthropic decay. The necropolis that neoliberalism has created in our cities—cities of the dead, dying and haunted—is that stuff that pervades our dark-dreams and constructs our present nightmares. In this aesthetic, that which threatens the psychological coherence of the self, is far too narrow an interpretation of today's texts. Monsters are now monstrous in the Derridian sense, as 'those that are appearing for the first time and consequently, are not yet recognised' (Derrida 1995, p. 386).

This section contemplates that the arrogance and the residue of human industrialisation in a variety of abandoned, residual, decaying and peripheral spaces of Gothic wastage. These are, necessarily, considered from a variety of perspectives, but all are connected to the waste and/or critical haunting that the Anthropocene has left. As a species, we have hollowed out the earth for our own devices and left its urban areas particularly afflicted by abandonment, inequity and loss. It is in these environments that we best encounter the Gothic aesthetic, but also philosophical issues around time. Dystopian dicachronity—the way language, etymology and concepts of anthropic dystopia have evolved over time—is significant in the interpretation of the Gothic in the age of the Anthropocene and its affects. If there is a new direction in Gothic Studies, it is one that is centred by the critical acceptance of the degeneration and ruin that the human has wrought in its epoch. Geography, Cultural Studies and the Digital Humanities are all seeing an appreciation of the abandoned, ruined and neglected spaces of modernity, particularly within the context of deindustrialisation. In these interrogations, as much as in Gothic story-worlds, ideologies of the Anthropocene proliferate. The New Media scholar Jussi Parikka, as a case in point, has written on a *Geology of Media* (2015) that is diachronically considered against the Anthropocene—in the same way I am arguing for Gothic Studies to be diachronically considered—such that its materiality is connected, temporally, to the earth's time. What is being suggested is a hybrid discourse in Gothic Studies—one that sits at the intersection of the aesthetic politics of the Anthropocene and tropes of the Gothic. This collection and, specifically, this section views the Anthropocene not as a theoretical problem that we need to push beyond, but as a hybrid sociocultural lens through which we can chart Gothic metanarratives.

The etymology of Anthropocene is from the Greek *Anthrōpos* (meaning human being) and *kainos* (meaning 'new') and it is used as word-forming element in geology to mean 'new age'. The concept has been criticised as purely ideological (Malm 2015) and some environmentalists on the political left suggest that 'Capitalocene' is a more historically appropriate term (Davies 2016) to use when commenting on the political and environmental aspects of post-capitalist society. At the same time, others suggest that the Anthropocene is overly focused on the human species, while ignoring systematic inequalities—such as imperialism and racism—that have also shaped the world (Anderson 2015).

The accepted timescale of the Anthropocene is, similarly, also a matter of debate. The age of the Anthropocene can stretch anywhere from: the invention of agriculture; or the expansion and colonisation by Europe; or the time of Capitalism that heralded the Industrial Revolution; or the first testing of the atomic bomb, through to today's trans-mediated digital revolution. Its ecological implications prompt questions about privacy, inequality, death and the ends of civilisations (Scranton 2013). And, the methods we use to historicise, remember, record, archive and inquire on the Anthropocene, are also implicated in its critique (Colebrook 2014). The digital revolution of the last thirty years has, arguably, increased popular awareness and anxieties over the Anthropocene's key tenets and queries and different Anthropocene's circulating in both the humanities and the sciences are subject to partisanship in the Academy as they are situated within the interests that each faction promotes. Thus, and for example, Tanya Krzywinska writing in this section writes not on the Anthropocene, but as the Chthulucene because it is specific to her subject field, yet the concepts discussed in her chapter are in line with our understanding of Gothic in the age of the Anthropocene. If the concept of the Anthropocene in Geology is relative to Earth Science time, yet the concept in critical thought is rooted in human-historical time, why then would we approach it in Gothic Studies in any other way that as part of a Gothic human-historical time continuum? 'Arguments in the humanities echo those in the geological sciences and are centred on the debate that distinguishes the epoch (and its conceptual traffic) as situated in either Earth time (planetary history) or Human time (world history)' argues Chakrabarty (2018, p. 6). However, arguments at the 2018 IGA conference, and elsewhere, have situated the debate in responsibility. As the blame seems to be largely systemic and specific to the system of capitalism complicated by post-war global economic systems, many argue that the epoch should more properly be called 'the Capitolocene' or 'the Econocene'. Others debate who used which designation first (Haraway 2015, ftn. 6). Regardless and irrespective of when we agree this epoch began, who is responsible for it, or how it should be called, what is clear is that we are most certainly *in it*, that it is unique in its relationships, and that our anxiety over it is the subject of most of our new Gothic narratives.

Although, aesthetics discourse has suffered from widespread suspicion post-Bourdieu and since the rise of populism, Gothic Studies has always depended on such discourse to reflect on cultural anxieties. The intersection of the Anthropocene and Gothic in the millennium has allowed for its continuance. That this diachronic is dystopian does not negate its value to the field. The chapters in this section reflect on these cultural anxieties and ethical dilemmas about the future—not just in our present historical time when our sense of security has become eroded in relation to our own identity, but also historically. Recalling that these anxieties have a lineage in the earlier forms of Gothic subjectivities, chapters in this section, ask how the ontological nature of 'catastrophe' is situated in our current Gothic tales. How do we negotiate the monster (biological, technical and ethical) that is our species? And what new landscapes are being drawn as a result?

There are significant shifts in Gothic signifiers in the *New Urban Gothic in the Anthropocene*. Monsters in the millennium are no longer the abject 'other'—they are each of us who are implicated in global or species collapse. Ghosts are no longer internalised and 'other'—they are ever-present, angry and prophetic. The double is no longer the assimilated 'other' in society—it is aesthetic device. The trope of the last man alive (Matheson 1954 [filmed as *The Last Man on Earth*, 1964; *The Omega Man*, 1971; and *I am Legend*, 2007]) is replacing these as the peripheral Gothic signifier. This 'Last Man' often represents the last man with a soul or morality—specifically drawn as an individual (the last) who does not have a relationship to the modes of production. A key recent text in this area comes from Thomas Nail (2016) who offers a new *Theory of the Border* and a new methodology for considering liminal states in our current social history. Specifically, Nail argues that there are now more borders and peripheral tribes of people today than ever before in history precisely because of globalisation and geo-media. These characters are appearing more often as the subjects of dark tales of peripherality. They are those that have been excised or exiled from society; the off-grid or the lost-tribes that the systems of production, consumption and globalisation have either left behind or wounded so integrally that they have reconstituted in wild-cultures. Similarly, whole cities that are below the radar of the Western metropolis (and the literature associated with them) are on the rise. Regions that, until recently, were being urbanised in the developing nations of China, India and South America are being drawn with the same sense of peripherality that the canonical locations of classical Gothic tales were in the Victorian period. The dialectics of core and periphery and the material cultures of these tensions are what are constructing our millennial Gothic narratives because the Gothic can and does account for society's combined and uneven development across disciplines and narrative forms. The Gothic is not a means of escape or escapism, it is a realistic social critique of combined and uneven development and residue of human violence—slow and cheap. The Gothic questions the systems and realism in and of society. It is not the political unconscious, rather it manipulates via textual fissures to extend those politics to dark and heightened tales of the peripheral and the centred; revenants and spectres; fetishistic deterioration and ruin; contagions and vortices; and bio-politics. All of these are aesthetics of the Anthropocene; all of these are narratives of remnants and residue that are central to this section. Together, they produce a sort of echo within the new urban Gothic form that we are seeing now.

Investigating millennial tales of the Anthropocene means highlighting the broader political and popular cultural contexts in which these narratives unfold, as well the complex ethical dilemmas they unmask. In general, and since the late modern period of posthumanism in the 1990s, there has been a repositioning of the human. Critical and fictional discourses now operate under the assumption (grounded in science, and presaged by Science Fiction since the 1950s) that we exist within an (ecological, economical, ethical) catastrophe of our own making and one that poses a real threat of extinction to our species. The underlying human experience of the sense of an ending, or as Žižek puts it, the anxiety of living in the end-times (2011), is prompting us to reconsider what human is, and what it means to dwell in the Anthropocene. Le Guin (1969) suggests that such narrative forms are speculative and less a prediction about the future than a thought-experiment about the present. The essays in this section are rooted in these speculations but draw on the tradition of the sublime to peak readers' interest. The potential future hazards and catastrophes of the Anthropocene are anxieties which have not necessarily materialised and may never materialise to the degree of extinction, but there is something potently sublime about living them out in tales such as these. These tales all hold a certain hauntology in common—the uncanny repetition of future paranoia. The extent of radicalism (and pessimism) with which the authors encounter in urban of the Anthropocene in these texts are broad, but all relate to Irmgard Emmelhainz's observation (2015) that the Anthropocene announces the collapse of the future through slow fragmentation. Or, as Roy Scranton succinctly put it at the beginning of his ironically titled book *Learning to Die in the Anthropocene*: 'We're fucked. The only question is how soon, and how badly' (2013).

References

Anderson, Ross. 2015. Nature Has Lost Its Meaning. *The Atlantic*, 30 November.

Bonneuil, Christophe, and Jean-Baptiste Fressoz. 2016. *The Shock of the Anthropocene. The Earth, History and Us*, trans. David Fernbach. London: Verso.

Broaks, Andrew. 2014. Do you Miss the Future. Mark Fisher, Interviewed. *Crack Magazine*, 12 September 2014. https://crackmagazine.net/article/long-reads/mark-fisher-interviewed/. Accessed 11 December 2019.

Chakrabarty, Dipesh. 2018. The Seventh *History and Theory* Lecture: Anthropocene Time. *History and Theory* 57 (1): 5–32.

Chase, Stuart. 1931. *The Nemesis of American Business*. London: Macmillan.

Colebrook, Claire. 2014. Archiviolithic: The Anthropocene and the Hetero-Archive. *The Derrida Today* 7 (1): 21–43.

Davies, Jeremy. 2016. *The Birth of the Anthropocene*. Oakland: University of California Press.

Derrida, Jacques. 1995. Passages—From Traumatism to Promise. In *Points. Interviews, 1974–1994*,, ed. Elizabeth Weber, 372–398. Stanford, CA: Stanford University Press.

Haraway, Donna. 2015. Anthropocene, Capitalocene, Plantationocene, Chthulucene: Making Kin. *Environmental Humanities* 6 (1): 159–165.

Irmgard, Emmelhainz. 2015. Conditions of Visuality Under the Anthropocene. *e-flux*, #63 (March). http://www.e-flux.com/journal/63/60882/conditions-of-visuality-under-the-anthropocene-and-images-of-the-anthropocene-to-come/. Accessed 11 December 2019.

Lawrence, Francis (dir). 2007. *I am Legend*.

Le Guin, Ursula. 1969. *The Left Hand of Darkness*. New York: Ace Books.

Malm, Andreas. 2015. The Anthropocene Myth. *Jacobin*. March. https://www.jacobinmag.com/2015/03/anthropocene-capitalism-climate-change/ (Accessed 11 December 2019).

Matheson, Richard. 1954. *I am Legend*. New York: Gold Medal.

Nail, Thomas. 2016. *Theory of the Border*. Oxford: Oxford University Press.

Parikka, Jussi. 2015. *A Geology of Media*. Minneapolis, MN: University of Minnesota Press.

Sagal, Boris (dir). 1971. *The Omega Man*.

Salkow, Sidney (dir). 1964. *The Last Man on Earth*.

Scranton, Roy. 2015. *Learning to Die in the Anthropocene*. San Francisco, CA: City Lights.

Shapiro, Stephen [in conversation with China Miéville]. 2008. Gothic Politics. *Gothic Studies* 10 (1): 61–70.

Žižek, Slavoj. 2011. *Living in the End Times*. London: Verso.

'A Weapon in the Cracks': Wasteways Between Worlds in the New Urban Gothic

Garth Sabo

The Gothic's utility for voicing and complicating cultural anxieties about external threats and social change has been well documented. Fred Botting (2004) sums up Gothic representations as 'a product of cultural anxieties about the nature of human identity, the stability of cultural formations, and processes of change' (p. 279). David Punter and Glennis Byron (2004) identify the Gothic as 'the symbolic site of a culture's discursive struggle to define and claim possession of the civilized, and to abject, or throw off, what is seen as other to that civilized self' (p. 5). This thematic analysis of Gothic conventions can be traced at least as far back as Eve Kosofsky Sedgwick's (1986) summary of Gothic narratives as an uncanny encounter between 'what's inside, what's outside, and what separates them' (p. 12).

Of course, grounding this volume in the new presumes some break or shift from the Gothic forms that preceded it. We must privilege arguments that highlight the evolution of the field since its origins in Kathleen Spencer's work with Dracula on the streets of London. Spencer historicizes the urban Gothic, finding its origins in the fantastic stories of the late

G. Sabo (✉)
Michigan State University, East Lansing, MI, USA
e-mail: sabogart@msu.edu

H.-G. Millette and R. Heholt (eds.),
The New Urban Gothic, Palgrave Gothic,
https://doi.org/10.1007/978-3-030-43777-0_6

nineteenth century. She also outlines the features of the genre, emphasizing the urban Gothic's modern setting and adoption of the discourse of empiricism. To make the new urban Gothic properly 'new' and sufficiently 'urban', we must remain attentive to the ways that the works within the genre do something more than merely continuing the work of their predecessors. This type of work must be predicated on finding, explaining and celebrating the city as a site of conflict between the empirical and the ethereal. Accordingly, I remain attentive to ways of differentiating this Gothic by the new forms its cities take, finding in the new urban Gothic an emergent form of the city rather than a mere return to interest in the conventions of haunted high rises and spooky street corners.

In this chapter, I suggest one way to theorize the shift I am describing. I look to two recent novels in the genre—China Miéville's uncanny detective thriller *The City and the City* (2009) and David Mitchell's ghost story *Slade House* (2015)—to argue that the new urban Gothic displays a profoundly different relation to rubbish. Trash destabilizes these novels, making it difficult, if not impossible, to imagine ontological spheres that can ignore or avoid contemporary ecological crises. In both cases, the irruptive force of things discarded punctures the classic Gothic tension between the world we know and the alien sphere that threatens it. Miéville and Mitchell craft a new urban Gothic that struggles to fit the inexorable momentum of material waste within existing generic boundaries.

Toxic Horror

The urban Gothic has long been steeped in filth. It cannot escape our notice that Kathleen Spencer turns to Mary Douglas, fairly considered the patron saint of waste studies, to formulate her thinking in 'Purity and Danger: *Dracula*, the urban Gothic, and the Late Victorian Degeneracy Crisis' (1992). As the title to her essay implies, Spencer draws extensively on Douglas' seminal *Purity and Danger: An Analysis of Concepts of Pollution and Taboo* (1966), in which Douglas presents her oft-cited dictum that dirt is 'matter out of place' (p. 44). Douglas' thesis that dirt does not exist outside cultural structures that rely on its presence to reify systems of control has been cited in scholarship far exceeding the anthropological context within which *Purity and Danger* was first published. Spencer invokes Douglas to stage the urban Gothic as an encounter between moral form and fears of formlessness sparked by rapid social change in *fin de siècle* city centres. 'Dracula', she argues, 'is a perfect example of the "formless" attacking form' (p. 220). She thus combines conventional understandings of the Gothic as a site of anxiety related to social change with dirt theory, which casts 'filth' as a moralizing term meant to rebuke challenges to the cultural status quo. The urban Gothic, as she sees it, does its work by deploying filth and other categories of waste to demarcate the lines between worlds.

Kathleen Spencer is not alone in her attention to the dirty underbelly of the urban Gothic. Her work on the urban Gothic builds on Seymour Rudin's use of the term to refer, more simply, to the surge in interest, in the 1960s, 1970s and 1980s, for creative works that transplant conventional Gothic horrors to twentieth-century urban settings. Rudin's essay 'The urban Gothic: From Transylvania to the South Bronx' (1984) predates Spencer's but prefers listing then-contemporary Gothic narratives to analyzing their themes. Nonetheless, he too invokes the trash-lined streets of the modern city to explain its suitability as a site for Gothic themes. Rudin focuses on texts and films with settings like 'the junk-choked wastes of the South Bronx' (p. 118), arguing that this and other 'analogous wastelands abroad ... have become the settings for the anti-human activities of the ancient horrors, settings that not only shelter the horrors but sometimes generate them, or their descendants' (p. 116). Rudin's urban Gothic similarly draws on the trash associated with its setting to mimic the eeriness of the haunted ruins and cobwebbed castles for which, according to him, earlier werewolf and vampire stories were known. His suggestion that these wastelands 'generate' horrors points to a natural affinity between trash and the Gothic.

In *Writing for an Endangered World: Literature, Culture and Environment in the U.S. and Beyond* (2001), Lawrence Buell paints this association between material waste and Gothic convention as inevitable rather than coincidental. Buell describes the 'expressed anxiety arising from perceived threat of environmental hazard due to chemical modification by human agency' as 'toxic discourse' (p. 31), in order to theorize how environmental and ecological concerns shape cultural imaginaries. Toxic discourse, moreover, 'readily montages into gothic' as its descriptive details accumulate (p. 42), meaning that, for Buell, Gothic conventions seem to be the logic endpoint for any narrative that portrays environmental factors as a potential threat or source of anxiety.

It is not surprising to see contemporary ecological concerns—the global sanitation crisis, the proliferation of material waste on both land

and sea, environmental toxicity, and climate change, to name a few—receive their Gothic due. As Andrew Smith and William Hughes note in their introduction to their recent *EcoGothic* collection (2013), '[d]ebates about climate change and environmental damage have been key issues on most industrialized countries' political agendas for some time ... The Gothic seems to be the form which is well placed to capture these anxieties' (p. 5). The range of essays curated for their collection speaks to the broad applicability that the ecocritical lens seems to have within Gothic studies. The authors of *EcoGothic* trace environmental and pastoral themes within the genre from its origins with *The Castle of Otranto* (1764) to contemporary works by authors as varied as Jack Kerouac, Margaret Atwood and Rana Dasgupta.

There does seem to be a limit, though, to what Buell calls 'toxic gothification' (p. 137). Despite chapters that address *flâneurs* and urban realist narratives, Buell does little to distinguish cities and their trash as unique sites for this type of discourse. This is due in part to the fact that, contra Spencer, Buell associates the Gothic primarily with rural settings. He sees the Gothic as 'another provenance of literary envisionment of the effects of environmental determinism in remote places' (p. 144), which suggests that cities are of a kind with every other haunted, polluted plane, valley or field. This may be well-suited to Buell's interest in the dynamics of situated thinking, but it unduly overlooks the extent to which the Gothic adapts to its setting. Similarly, Smith and Hughes' attention to the ecological issues that have informed the Gothic 'for some time' presumes a flat continuity between the environmental pressures of the Victorian era and the ecological crises to which the texts of today respond.

All of this is meant to suggest, simply, a collision course between the ecological and urban Gothics that I find eminently productive for reflecting on the current moment. I propose trash as a useful lens through which to appraise this friction. After all, like the Gothic, trash is *by its nature* caught up in the past. William Viney's *Waste: A Philosophy of Things* (2014) suggests that with 'our recognition of waste comes an acknowledgment of time's passing, its power to organize notions of wearing, decay, transience and dissolution and its power to expose that organizing function' (p. 4). Michael Thompson's seminal *Rubbish Theory: The Creation and Destruction of Value*, similarly, argues that the category of rubbish serves to mediate competing conceptions of finite and infinite material lifespans, meaning that trash is always already bound up in thinking about the past and its possible relations to the future. In the new urban Gothic, as I will show, the haunting temporality of trash starts to unravel the ghost stories we used to know.

'I'M MY OWN LEFTOVERS': UNHAUNTING *SLADE HOUSE*

Slade House is a strange book. Though published in 2015 as something of a prequel to David Mitchell's *The Bone Clocks* (2014), *Slade House* has been largely overlooked in favour of its more ambitious, genre-defying companion text. By comparison, in fact, Mitchell's take on a classic ghost story seems almost *too* conventional. The novel centres on the spectral twins Norah and Jonah Grayer who, having learned to thwart death, must feed on the soul of psychically gifted individuals once every nine years in order to maintain their immortality. The twins' mystic energies keep their bodies and the titular London mansion frozen in time 'a few minutes after 11 P.M. on Saturday, 27 October, 1934' (p. 180), even as the majority of the plot unfolds over the course of decades from 1979 to 2015. The Grayers lure their victims through portals into their psychic domain and trick them into eating an offered food item and thereby forfeiting their souls. Mitchell sends several curious Londoners to a gruesome end in nine-year increments until the ghostly pair is finally thwarted by Marinus, a rival immortal being and hero of *The Bone Clocks*.

At first glance, *Slade House* abides by Spencer's rules for the urban Gothic with little alteration. The tale's urban setting is emphasized repeatedly throughout the novel, with the early assertion that Slade is 'only the Grayers' town residence. Their proper home's in Cambridgeshire' (p. 6) serving well as a nod to the genre's origins outside the city proper. Mitchell's explanation of the Grayers' immortality seems to follow the urban Gothic's discourse of empiricism as well. Expository pseudoscientific monologues serve as something of an aperitif for the Grayers' consumption of souls, as when Norah chides her brother for his carelessness during their latest hunt, reminding him that

> The operandi works pro*vi*ded our birth-bodies remain here in the lacuna, freeze-dried against world-time, anchoring our souls in life. The operandi works pro*vi*ded we recharge the lacuna every nine years by luring a gullible Engifted into a suitable orison. The operandi works pro*vi*ded our guests can be duped, banjaxed and drawn into the lacuna. Too many *provideds*, Jonah. Yes, our luck's held so far. It can't hold forever, and it won't. (p. 79)[1]

The Grayers continue to define their terms to a later quarry. Speaking to their quarry through the possessed body of Fred Pink through a technique they call 'suasioning' (p. 176), the twins explain that a lacuna is 'small space that's immune to time' (176), an orison is a 'kind of reality bubble' (p. 177), and banjax is 'a chemical that shrivels the cord fastening the soul to the body, so it can be extracted just before death' (p. 177). The whole system 'runs off psychovoltage' (p.177), which all but completes the twins' painstaking efforts to couch their uncanny abilities in empirical terms.

The extent to which this explanatory babble punctures the narrative, however, suggests that Mitchell is more interested in drawing attention to the generic pressure to include such discourse than he is in integrating it uncritically into the text. In *Contemporary Gothic* (2006), Catherine Spooner suggests this kind of self-consciousness as a key feature of the genre's adaptation to the twenty-first century. Spooner argues that the psychoanalytical, Marxist and feminist revolutions of the twentieth century have imbued the new Gothic with 'a sexual and political self-consciousness unavailable to the earliest Gothic novelists' (p. 23). Reading Mitchell, this seems to be particularly true. His observance of convention in *Slade House* toes the line between homage and parody, with the end result being a ghost story where the long history of the Gothic emerges as the true spectral presence haunting the text.

The Grayers are uniquely beset upon by threats, despite their ostensible position as the villains at the heart of the Gothic narrative. For all their uncanny power, the spectral twins' positions are increasingly tenuous throughout the novel. Norah is particularly aware that the world is changing, and the place in it for undying psychic ghouls is shrinking. Norah laments that every return to her birth-body, though necessary to feed on a new soul's psychovoltage, leaves her feeling 'more enfeebled' (p. 33). Given that the twins' bodies are insulated against time within the lacuna, this insinuation of bodily decay suggests a preoccupation with the Gothic's capacity to balance certain tenets of empiricism with the fantastic. Mitchell invites paradox into his haunted house by depicting repeat visits to a timeless place. Time may not pass in the lacuna, but Norah and Jonah do, between astral projections into the world and feasting on souls by candlelight. Mitchell's exhaustive specification of the mechanics of their operandi elsewhere in the text casts the impossibility of the atemporality at the heart of it into sharp relief. Late in the text, Jonah, paralyzed in a climactic surprise attack and unable to project himself past the lacuna's borders, complains that '[t]hirteen months is thirteen eternities if you're stuck in a lacuna' (p. 206). The slippage between finite and indefinite temporalities here speaks directly to the problems of fitting a truly sublime space, such as the room out of time, into a narrative purporting to abide by empirical rules.

As Norah's distaste for her enfeebled embodiment suggests, these problems of Gothic chronology within *Slade House* revolve around questions of waste, inexorable and omnipresent throughout the novel. The Grayers are threatened, and ultimately undone, by the lingering materiality of their victims. Time and again, Mitchell's attention falls to the fragmentary pieces that gradually accumulate and challenge the twins' mastery of the overlapping psychic and physical spaces. The ghosts that haunt Slade House are themselves haunted by the traces of their former victims, which manifest as spectral afterimages that wander the house with eerie warnings for new guests to escape while they can. Each new hunt is haunted by the remnants of the preceding victim, leading the more cautious Norah to worry that the 'time will come when a "harmless afterimage" will sabotage an Open Day' (p. 33).

The terms use to describe these spectral traces emphasize their ephemerality and materiality in equal measure. After his soul is harvested, the detective Gordon Edmonds is reduced to 'just the residue' of his former self (p. 82). The ghost of his predecessor describes himself as 'my own leftovers' (p. 70). The Grayers, speaking through their possessed interlocutor, offer the clearest connection between their operandi and the production of material waste. Once their victims' souls have been devoured, the bodies that remain are 'chucked into the gap between the orison and our world. Like bin bags down a garbage chute' (p. 180). When the afterimages of souls return from this garbage space between world to wreak their vengeance, it clearly stages a conflict between the Grayers and the irruptive power of contemporary waste politics, which push against the clear separation of worlds that has previously been so intrinsic to the Gothic's cultivation of the uncanny.

This is far from the only moment or way that trash threatens spectral immortality in *Slade House*. Jonah, speaking through Fred Pink, portrays the accumulative aspect of materiality as inimical to eternity. Asked why the Grayers kept their discovery of immortality to themselves instead of sharing it with the world, Jonah's answer is simple:

> Give it twenty, thirty, fifty years, there'd be thirty, forty, a hundred billion human beings eating up our godforsaken world. We'd be drowning in our own shit even as we fought each other for the last Pot Noodle in the last supermarket. See? Either way you lose. If you're smart enough to discover immortality, you're smart enough to ensure your own supply and keep very very *very* shtum indeed. (pp. 175–176)

The image of a world overflowing with immortal beings drowning in their own excrement is certainly evocative here. Moreover, it establishes a clear opposition between the atemporality of a life unending and the inevitability of waste as a component of material embodiment. The central conflict of *Slade House* unfolds less between the spectres and their potential victims than between the incompatible conventions of intelligibility that compose such literary hauntings on the one hand and the lingering dread of material waste on the other. Thus, Mitchell's ghost story contests the inexhaustible demands material waste places on our attention. If the Gothic has traditionally been built around the horror of an alien world from which escape is uncertain or impossible, the new urban Gothic seems more attuned to the ways that our own material crises inevitably leak into these new narrative planes.

In her excellent *Excrement in the Late Middle Ages* (2008), Susan Signe Morrison coins the term 'wasteways' to denote the material and conceptual paths cultures use to order their dejecta. Though Morrison's thesis mainly concerns the circulation of human excrement, she does not exclude rubbish and other forms of material waste from her wasteways; we need not either. The field of waste studies is predicated on paying attention to waste, in all its guises, as a cultural force. Adopting Morrison's wasteways offers a useful way to conceptualize the irruptive power of trash as a function of its momentum as it moves within representations of our world. David Mitchell's besieged ghosts suggest that, within the new urban Gothic, wasteways begin to cross and even challenge boundaries within and between our worlds. Within *Slade House*, these wasteways line the gaps between worlds, a hidden threat lurking in the '*cracks they throw the scraps down*' (p. 70) that allows our castoffs to climb back into our field of view and displace those sublime, impossible horrors for which the Gothic has been known. There is, Mitchell shows, much to be found in these garbage-filled gaps between worlds. As one of his afterimages suggests, '*You may find a weapon in the cracks*' (p. 70). This, along with Norah's concern over the so-called harmless afterimages, turns out to be prophetic when a later shade returns from the dumping ground between worlds to stab Jonah through the throat and thwart the twins' penultimate soul-harvesting ceremony. *Slade House*, I argue, displays a willingness for the new urban Gothic text to aim such weapons at the Gothic itself.

Mitchell is careful to claim this new trajectory of waste as a recent development within the Gothic; in other words, *Slade House* aligns the 'new' of the new urban Gothic with this burgeoning conflict between the supernatural and imminently material. As the novel draws to an end, the Grayer twins meet their match in Iris Marinus-Fenby, a benevolent immortal known as a Horologist and key figure in *The Bone Clocks*. Marinus' longevity works not by thwarting death but by dying and being reborn in an eternal cycle. Given the potentially endless trail of corpses this form of immortality would imply, it stands out that the conflict driving this novel and its companion text centres around balancing the supernatural with an awareness of the persisting material traces of existence. Before dispatching the parasitic Grayers, Marinus explains their undoing:

> Your 'crematorium' disposes of bodies well enough, but inorganic matter falls through the cracks. In the old days it hardly mattered – a button here, a hair clip there, but in this century [...] angels really do fit onto pinheads, and the lives of the multitudes inside a memory stick. (p. 227)

This last is a reference to the digital tape recorder the twins' previous victim had used to record their self-satisfied monologues, which has fallen, as it were, through the cracks of their place out of time and given Marinus the means by which to destroy them. This final section of the text is dated 2015, meaning that her quip about 'this century' historicizes this new era to the millennium fairly cleanly. Marinus' casual dismissal of 'the old days' implies a substantive break from the long history of the Gothic when the insistent materiality of fragments and castoffs could abide inaction. As I will show in the next section, the cracks between worlds, increasingly seen as the plane of trash, become a powerful weapon for the work of the new urban Gothic.

'It's just Rubbish, and It Drifts Across Borders': The Cracks Between China Miéville's Cities

David Mitchell is, of course, far from the only contemporary writer interested in the Gothic potential of the cracks between worlds. China Miéville's celebrated novel *The City and the City* imagines a cityscape shared by two cities, Besźel and Ul Qoma, whose inhabitants are legally required to unsee the people, vehicles and buildings of the other city even if their paths cross in crosshatched streets—that is, areas of the city that exist in both Besźel and Ul Qoma. Miéville's protagonist, the Besź detective Tyador Borlú, must investigate the murder of a student killed in Ul Qoma and dumped in Besźel. This constitutes a serious transgression of the cities' hallowed rules of separated coexistence, and seems to clearly invoke Breach, the unseen, uncanny force that maintains the border between the two cities. When Borlú's attempts to hand the case off to Breach are stymied, he pursues the investigation himself, and in the process unearths rumours of the third city, Orciny, hidden in plain sight and lurking in the cracks between Besźel and Ul Qoma.

The Gothic foundations of the text are fairly transparent. Inspector Borlú and his partners Corwi and Dhatt continue the tradition of the canny, hard-drinking detective illuminating the macabre with deductive reasoning and intellectual prowess. Of more interest to this chapter, however, is the way Miéville casts the residents of Besźel and Ul Qoma as unknowable Others to their neighbours. The Besź and Ul Qomans, who are neighbours grosstopically,[2] but remain 'a city away' from one another (p. 253), reciprocally haunt one another. The strict rules of Breach obligate a person in Besźel to recognize and react only to those elements of the surrounding area that are also in Besźel. If an Ul Qoman car backfires, the Besź can't hear it; the smells of Ul Qoman food can't make the Besź hungry. The residents of both cities are conditioned from birth to avoid any outward signs of recognition of the spectral others that share their streets, for fear of Breach.

This does not, of course, mean that the two cities remain perfectly alienated from one another. Borlú takes a brief, perverse pleasure in a split-second of eye contact with an Ul Qoman passing by on a train that, for Borlú, legally does not exist, and surreptitious glances at attractive passersby in the other city's streets seem to be a universally indulged gray area in the rules of Breach. These present absences leave the streets of Besźel and Ul Qoma materially haunted by the ghostly inhabitants of their neighbour city. Together Beszél and Ul Qoma invoke the ethereal presence of the spectral other, which Miéville juxtaposes against the discourses of legality and urban materiality.

As these two cities reciprocally haunt one another to activate the uncanny within *The City and the City*, the urban ghost story is complicated by Orciny, the third city hidden in the cracks between. Even the suggestion of Orciny manifests a threat to both cities. The investigative trail that points again and again to the third city suggests 'systemic transgression, secret para-rules, a parasite city where there should be nothing but nothing, nothing but Breach' (p. 297). The novel reaches its crescendo when Bowden, author of a discredited treatise on Orciny, takes advantage of anarchy within both cities to perform a quantum state of citizenship where neither Besź nor Ul Qoman can determine whether he dwells in their city or their neighbour's. Bowden deploys his academic expertise on the people of both cities to cast himself into the third space of the neither-nor, 'mediat[ing] those million unnoticed mannerisms that marked out civic specificity, to refuse either aggregate of behaviours' (p. 354). The result is uncanny and grotesque. Neither Corwi in Beszél nor Dhatt in Ul Qoma knows how or whether to interact with 'Schrödinger's pedestrian' (p. 352), whose facial expressions and gestures effect an unsettling amalgam of the conventions in both cities. The true horror of *The City and the City* is decidedly non-binary, an uncertain phantom that disrupts the conventional hauntings accepted as the status quo throughout the novel.

Miéville's characters struggle to differentiate Orciny from Breach, given their shared liminalities. Both entities endanger the inhabitants of Beszél and Ul Qoma by challenging the either-or ontology of the Gothic play of binaries. Orciny introduces the horror of the neither-nor. Once Borlú enters Breach, he discovers the both-and or, as he puts it, a 'nowhere-both' (p. 306), interstice and site in equal measure. In this, he rebuffs the earlier suspicion that 'Orciny is the name Breach calls itself' (p. 254). Breach is not in a space different from Beszél or Ul Qoma; it is simply 'in both' (p. 304). Thus whereas Orciny threatens with negation, the excess of Breach punctures the boundaries between the cities, showing them to be permeable. Indeed, each city's aloofness to the other is predicated upon their mutual openness to Breach. This is the type of conventional self-consciousness that Spooner attributes to the contemporary Gothic, in which the rules that construct the genre stage a conflict between discrete, incompatible worlds while also structuring those worlds from the third, liminal position.

Breach does not wield this power of doubled presence alone; in fact, for all the extraordinary authority they wield, the avatars of Breach share their strength with some remarkably mundane objects within the city. Borlú elicits a smile from Dhatt by comparing the 'drizzle' in Ul Qoma with the '*real* rain' they get in Besźel, noting that this type of playful reference to the cities' grosstopic imbrication is recognizable as an 'old joke' in both cities (p. 237). This is in keeping with the Besź proverb that 'Rain and woodsmoke live in both cities' (p. 65). This would suggest that Breach lives in the nowhere-both alongside other natural elements like weather conditions, except that, as Borlú notes, the same is also said about 'even rubbish, sewage, and, spoken by the daring, pigeons or wolves' (p. 65).

Here, trash erupts past the boundaries between worlds. In fact, the offal of the streets appears to openly flout the rules of separation between Besźel and Ul Qoma. The residents of both cities are under no obligation to unsee litter, because trash exists doubly in both places:

> Only rubbish is an exception, when it is old enough. Lying across crosshatched pavement or gusted into an alter area from where it was dropped, it starts as protub, but after a long enough time for it to fade and the Illitan or Besź script to be obscured by filth and bleached by light, and when it coagulates with other rubbish, including rubbish from the other city, it's just rubbish, and it drifts across borders, like fog, rain and smoke. (p. 80)

Miéville plays with time on several scales here, first noting that rubbish attains this duality only 'when it is old enough'. This is complicated, of course, with his deployment of 'other rubbish' and then 'just rubbish'. To be 'just rubbish' suggests the shedding of extraneous details, whereas coagulating with 'other rubbish' reiterates the always-already effect of rubbish as an identifying category. For rubbish to become 'just rubbish' when it joins with other rubbish presumes a twisted chronology in which the past is simultaneously refuted and reified, made manifest in waste's doubled interruptive presence in both cities.

And as it 'drifts across borders', trash reminds us that wasteways are grosstopic structures. Miéville suggests that, like a ghost, material waste passes freely between spheres of existence that purport to be closed and impenetrable. Objects remain protub only so long as their cultural markings remain legible. The nuances of each city's language are lost beneath an accumulation of grime and the fading effects of exposure, just as, elsewhere in the novel, the elaborate politics of colour in each city are shown to be remarkably fragile. Borlú is particularly attentive to the legal distinctions between shades of blue that are deemed fit for one city but not the other. He highlights one 'shade called Besźel Blue, one of the colours illegal in Ul Qoma' (p. 64), and later finds that even simple environmental factors can reduce the subtle distinctions between each city's legal hues. On one particularly freezing day, cold mut[es] Ul Qoma's colours to everyday shades' (p. 189). Miéville here effectively demonstrates the continuous effort needed to maintain the markers of specific cultural value and the ease with which an object can tumble into both cities as waste. As inattention and exposure obscure the cultural markers of an Ul Qoman or Besźel origin, material objects erupt out of the sphere of cultural significance and reveal how thin the veil of commodity value truly is.

Rubbish and Breach dwell together in the liminal spaces of *The City and the City*, though Breach hides while trash announces its presence in the novel's gutters and empty lots. This spectral trash pushes aside existing Gothic structures, displacing the conventions of haunting with its own insistently material presence. In fact, we increasingly find trash performing this way in works of the new urban Gothic. It pushes in from the cracks between worlds, delineating each and threatening both.

Spectrality in the Dumps

In *Specters of Marx: The State of the Debt, the Work of Mourning and the New International* (1994), Jacques Derrida coins the term 'hauntology' to refer to the spectral presence of commodities. Commodity-form and use-value are not the same, but they do mutually haunt one another. Though the commodity is not and cannot be present in the use-value of the wooden table, it 'affects and bereaves it in advance, like the ghost it will become' (p. 201). Hauntology denotes this spectral presence at the heart of all ontology, in which all material presence invokes the spectre of absence. As the examples of Mitchell and Miéville remind us, the commodity is not the only object form that haunts in this way. Trash, with its temporality of decay and irruptive power embodies a similar spectrality.

The new urban Gothic seems to be interested in Derridean hauntology as the source of new ghost stories. Derrida remarks that 'every period has its ghosts (and we have ours), its own experience, its own medium, and its proper hauntological media' (p. 222). Given the way these novels stage disruptive encounters with the litter and leavings of the contemporary city, I have argued here that the new urban Gothic may be the proper hauntological medium for the horrors of the waste crisis. For while trash in both *Slade House* and *The City and the City* exorcises certain ghostly threats, it does so only by dint of its ability to permeate, to leak, to constrain and to control. Its threatening presence is invoked even in its increasingly rare absence. The new urban Gothic stages the omnipresence and narrative omnipotence of waste as a conflict, already won, between the ethereal hauntings of Gothic tradition and the raw de(con)structive power of trash. In so doing, it substitutes the insistently material for the supernatural, displacing the uncanny with the garbage can.

Notes

1. All italicized quotations in this section are original to Mitchell.
2. Miéville coins the phrase 'grosstopic' to denote the physical space shared by the cities. Buildings in Besźel can only be next to ones in Ul Qoma grosstopically. Areas that belong explicitly to the other city are 'alter' spaces, and areas shared by both cities are 'cross-hatched.' Alter objects that must be unseen and casually avoided are known as protubs, an abbreviation for 'protuberance'.

References

Botting, F. 2004. Aftergothic: Consumption, Machines, and Black Holes. In *The Cambridge Companion to Gothic Fiction*, ed. J. Hogle, 277–300. Cambridge: Cambridge University Press.

Buell, L. 2001. *Writing for an Endangered World: Literature, Culture, and Environment in the U.S. and Beyond*. Cambridge: Harvard University Press.

Derrida, J. 1994. *Specters of Marx: The State of the Debt, the Work of Mourning and the New International*. New York: Routledge.

Douglas, M. 1966. *Purity and Danger: An Analysis of Concept of Pollution and Taboo*. New York: Routledge.

Miéville, C. 2009. *The City and the City*. New York: Ballantine.

Mitchell, D. 2015. *Slade House*. New York: Random House.

Morrison, S. 2008. *Excrement in the Late Middle Ages: Sacred Filth and Chaucer's Fecopoetics*. New York: Palgrave Macmillan.
Punter, D., and G. Byron. 2004. *The Gothic*. Oxford: Blackwell.
Rudin, S. 1984. The Urban Gothic: From Transylvania to the South Bronx. *Extrapolation* 25 (2): 115–126.
Sedgwick, E. 1986. *The Coherence of Gothic Conventions*. New York: Methuen.
Smith, A., and W. Hughes. 2013. Introduction: Defining the EcoGothic. In *EcoGothic*, ed. A. Smith and W. Hughes, 1–14. Manchester: Manchester University Press.
Spencer, K. 1992. Purity and Danger: *Dracula*, the Urban Gothic, and the Late Victorian Degeneracy Crisis. *ELH* 59 (1): 197–225.
Spooner, C. 2006. *Contemporary Gothic*. London: Reaktion Books.

The City and the Underground in *Metro 2033* and *Metro: Last Light*

Madelon Hoedt

It begins with a radio message: 'Rockets are launched. This is the end'. He runs out of his apartment, onto the street, shouting a brief message to anyone he comes across, anyone who will listen: 'Go down to the Metro!' Sirens go off in the distance, followed by a blast, an explosion that is felt more than heard. The man hurries on, dodging between people, between cars, a mad scramble of bodies, possessions and panic. The Metro, by contrast, is empty, sterile, magnificent, the military guarding its doors. They are only able to admit so many into the halls of the people's palace and warning shots are fired in an attempt to control the throng as the doors, ornately decorated with symbols of the Soviet Union, are closed before those who will remain above ground. Whether those inside have reached safety or a different kind of grave, one can only guess. The captions of the trailer for the video game *Metro: Last Light* (4A Games, 2013)[1] tells viewers that these events date back to Moscow in 2013. Now, twenty years later, the story continues as *Metro 2033*.

M. Hoedt (✉)
Department of Drama, Theatre and Performance,
University of Huddersfield, Huddersfield, UK
e-mail: m.e.hoedt@hud.ac.uk

H.-G. Millette and R. Heholt (eds.),
The New Urban Gothic, Palgrave Gothic,
https://doi.org/10.1007/978-3-030-43777-0_7

Set in Moscow following the nuclear attack, this expansive transmedia franchise started with the original *Metro 2033* novel, which was posted online by author Dmitry Glukhovsky in 2002. Gaining in popularity, the title received a physical release in Russia in 2005 and an English translation in 2010, and further elements have been added to the series since. At the core of these are the trilogy of novels, consisting of *Metro 2033*, *Metro 2034* and *Metro 2035*, written by Glukhovsky himself, but at the time of writing, over 50 further novels set in the Metro universe exist, penned by other authors.[2] In addition, other media have been released, many of which take the form of games: there is a *Metro 2033* board game, at least two Russian language online games, and, the focus of this chapter, the two video games released by Ukrainian developer 4A Games, titled *Metro 2033* (2010) and *Metro: Last Light* (2013).[3] Each takes the 2013 nuclear disaster as their starting point and allows players to explore the ruins of Moscow as they exist after the attack. Muscovites have been forced to abandon the city and move underground in order to escape radiation poisoning and mutated creatures, turning the iconic Metro system into their home. Each Metro station has become a base for different factions and alliances, ideologies and governmental structures—a functioning microcosm of society, underground. The *Metro* universe has proven particularly evocative for readers, where the centrality of location and of the spaces in which the narrative plays out are framed clearly through the choice of its underground setting with its inherent symbolism. Yet this exploration is ultimately story-driven: in both games, players will follow protagonist Artyom as he journeys through the Metro and traverses the hostile environments of the new Moscow, both above and below ground. Using a first-person perspective within the game experience, the *Metro* franchise places players in a similar position to Artyom, mirroring his encounters with the ruined cityscape and its reversal of underground and overground spaces. The aim of this chapter is to explore this aspect of the two video games through a discussion of the visualization of and commentaries on this new landscape. Starting with an exploration of the city of Moscow as construct and symbol, the chapter will then move onto a discussion of the game world and its representation of urban spaces. The work of Augé and Tuan on (non) place and space will be used to explore this new hierarchy, and I will conclude with a reframing of post-apocalyptic Moscow as an example of Foucault's heterotopic space, holding its contradictions within itself.[4]

Before focusing on the game texts, however, it is important to understand the context of Glukhovsky's reimagining of the city. The move from Moscow overground to Moscow underground presents an interesting problem, as both sides of the city have swapped places, copying and doubling the function of the other. In doing so, the promise of what might be lurking underneath is realized and its relation to ground-level subverted, leaving a space that is ready to be (re)discovered by its inhabitants. Ordinarily, the Metro counts as a method of transport, a place to traverse through, where the tunnels themselves are largely unseen, hidden by the bright lights inside the train cars. The underground is familiar through daily use by so many commuters, but its tunnels are not truly known: the dark passages and endless expanses of rails could hide any number of dangers. The imagined 2013 nuclear attack changes this dynamic: it forces the population into its depth, abandoning their routine existence in Moscow's streets, where the Metro must now become a new home, a new order. A deeper knowledge of the transport system becomes a necessity as Muscovites try to build a new life below ground.

Present in the novels, this notion is emphasized within the videogames, which require players to take on the role of Artyom and to navigate both sides of the city. In using video games as the focus of this chapter, as opposed to, for example, the novels that form part of the franchise, it is assumed that they offer a particular experience for audiences. Although I wish to avoid a discussion of immersion and interactivity, and what these add to a playable, as opposed to a reading, experience, I do wish to briefly discuss the world-building that can occur within the medium of digital games. Video games allow for the imaginary spaces created in the text of the novels to be realized in such a way as to allow players to reside within them. Defined by Totten as a designed space (2014, p. xxii), the level design in games creates a possibility space for players to move around inside the fictional world, and it is Jenkins who reminds us of the importance and potential of this process: 'Game designers don't just tell stories; they design worlds and sculpt spaces' (2004, p. 121). The designed space of the video game presents a realized fictional space, which exists both visually and spatially, where the levels can take on at least part of the burden of story and exposition: 'The environment provides narrative context' (Worch and Smith 2010, slide 13). Players can enter into and move through the digital representation of a world, and this sense of place can offer them a particular insight into these spaces.

In the case of the *Metro* games, this is primarily a (re)presentation of the urban landscape of Moscow as it exists above and below ground, including encounters with some existing spaces and recognizable landmarks. As such, an interpretation of both games as employing an urban Gothic setting is apt, but within the context proposed here, the use of the term needs to be reconsidered. Mighall defines the concept in terms of opposition, where the concept of urban Gothic 'depicted what the city (civilisation) banished or refused to acknowledge, except in the form of thrilling fictions' (2007, p. 54). Soon, however, the promise of modernity connects to a darker side of the city which, as Warwick notes, 'is seen as uncanny, constructed by people yet unknowable by the individual' (1998, p. 288). It is a space of progress, of people, yet simultaneously 'a place of ruins, paradoxically always new but always decaying' (Warwick 1998, p. 288) in which reside 'a concentration of memories and historical associations' (Mighall 2007, p. 57). This is perhaps best summed up by Mighall when he states that the urban Gothic is 'not just a Gothic in the city, it is a Gothic *of* the city. Its terrors derive from situations peculiar to, and firmly located within, the urban experience' (1999, p. 30; author's emphasis). Arguably, in the *Metro* universe, such an 'urban experience' is lacking after the blast, as the populace have moved from the city streets into its tunnels. The movement in *Metro* from overground to underground presents a particular shift in the experience and understanding of the urban Gothic space.

Leaving aside their fictional representation, the meanings inherent in the games' use of real-life settings and their connotations need to be considered. Not only are the proposed oppositions between above and below ground of importance; Moscow's status as a city and a symbol needs to be taken into consideration. In much of Urban Gothic scholarship, cities like London take centre-stage, with academics discussing the different ways in which texts have mapped these spaces (for example, Luckhurst 2002; Phillips and Witchard 2010). This includes both fictional settings and the location of production; one can comfortably speak of an English Gothic fiction to which, as Cornwell notes, 'exotic' European elements may have been added (for example, German and Italian settings) (1999, p. 9). Although referenced in scholarship on the Global Gothic, exploring the ways in which Gothic traditions have been located in national and transnational contexts (see Byron 2012, 2015), little work is available on a Russian or wider Eastern European Gothic tradition. Indeed, its existence is dispersed: Cornwell notes that Russian works are more likely to be categorized as romanticism or expressions of the fantastic, drawing on a lineage of medieval themes and folktales (1999, pp. 3–4), and Bowers has pointed to the presence of Gothic motifs in realist works (2013, 2017). Gothic works in Russia appear few and far between, both in imported works or texts in translation, and in novels by Russian authors. Despite this apparent lack of canon, Cornwell does point out that 'Gothic appurtenances continued to resurface in Russian literature', going so far as to identify a subset of 'St. Petersburg Gothic' (1999, p. 19); however, he fails to specify what exactly this might mean, or what themes and motifs might inform this Russian mode.

Cornwell does not bestow a similar accolade on Moscow and the city is hardly mentioned in his book. One might think that the turbulent history and political strife seen in Russia's current capital would provide inspiration for novels with a darker tone, and although the texts are not described as Gothic, Moscow, too, has been the site of significant literary contributions. The realist portrayal of the city in Leo Tolstoy's *Anna Karenina* (1877) is contrasted by the magic realism of *The Master and Margarita* (1967) by Mikhail Bulgakov and, much later, the satirical dystopia in Vladimir Sorokin's *Day of the Oprichnik* (2006), to give a few examples. The works mentioned here each provide their take on the city: Tolstoy's portrayal of imperial Russian society; Bulgakov's parable of the Devil's visit to Moscow and the havoc he wreaks; and Sorokin's reimagining of a Tsarist Russia in the year 2037, each carrying themes of ongoing history and political upheaval within their texts. Yet I argue that the symbolic status of Moscow and the myths that surround the city are best exemplified in Anton Chekhov's 1900 play *Three Sisters* in which the titular sisters, now residing in the countryside, display a nostalgia and longing for the memories of Moscow, evidenced in the stories they tell each other of their city. Here, as in the other novels, Moscow is the creator of its own tales, a curious mixture of realism and mythmaking, of old attitudes and new developments, a fictional construct rather than an actual brick-and-mortar urban sprawl.

The Metro holds a similar status, an architectural marvel and a feat of engineering, and it is important to briefly explore this symbolic representation of the transport system before moving on to a discussion of the games. Described by Meuser and Martovitskaya as 'one of the most beautiful metro systems in the world' (2016, p. 7), the Moscow Metro cannot be disconnected from what it represents. Opened in 1935, the construction of the network has been ongoing for nearly a hundred years, and spans as many years of the city's past. Many historical events are reflected in its elaborate design and decorations, where the network is regarded as a 'venerable art museum', and its stations as 'cathedrals of the underworld' which 'still fill Muscovites with pride and respect' (Meuser and Martovitskaya 2016, p. 6). Kuznetsov comments on the connection between the Moscow Metro and political ideology, where 'the metro was from the start intended not as a purely utilitarian piece of the city's transport infrastructure, but as a unique architectural structure reflecting the development of both city and country' (2016, p. 36). More specifically, Kuznetsov notes that 'over the course of the greater part of its history the architecture of the Moscow metro has carried a serious ideological burden: it has had to serve as a vivid incarnation of the superiority of Socialist over Capitalist society' (2016, p. 41). This political symbolism in turn impacted on the design of the stations, aimed to be 'people's palaces' (Kuznetsov 2016, p. 36): Zmeul describes how 'the architecture of Moscow's metro went from severe functionality to temple-like grandeur' (2016, p. 76) in which feats of engineering made way for artistic expression and decoration. Examples of this can be found in numerous stations, such as the 76 bronze statues at Ploshhad' Revoljucii (Revolution Square), depicting soldiers, farmers, and factory workers, and the ceiling art at Komsomolskaya (named after the Komsomol, the Soviet youth organisation, image 1).

The icons of the Soviet regime, the figures of citizens and soldiers and the emblems of the Communist party endure within the tunnels of these, and many other, stations. Later additions to the Metro system, specifically those constructed at the time of the Great Patriotic War (WWII), represent spaces which were to be 'the first monuments to the upcoming victory' and which later displayed 'the triumph of the Soviet people', resulting in an architectural style which Zmeul describes as 'not so much palatial as almost temple-like and sacral' (2016, pp. 148, 151). Framed by this mixture of politics and design, it is impossible to disconnect the Moscow Metro from its associations with the Soviet regime and ideology. In addition to depictions of the accomplishments of the USSR, the ornamentation of the stations employs a wealth of materials: halls are decorated with paintings, mosaics, frescos, bronze statues; a lavish décor and rich colours. Over the course of its construction the Metro 'went on to organically absorb the ideological and stylistic qualities of each successive epoch, becoming a unique encyclopaedia of Soviet, post-Soviet, and new-Russian architecture' (Meuser and Martovitskaya 2016,

Image 1 Chandelier and ceiling detail at Komsomolskaya metro station

p. 7). In many ways, the Moscow Metro now stands as a demonstration of the might of Mother Russia in all its incarnations.

It is against this backdrop of symbolism and ideology that the games find their evocative setting. Both *Metro 2033* and *Metro: Last Light* include similar game mechanics, drawing on conventions of first person shooter (FPS) and survival horror games. Players are sent into a world where they will be confronted with a number of different enemies (both human and supernatural). In order to defend themselves, they have access to a variety of firearms, as well as some additional tools (grenades, knife, lighter) to help them navigate the dark tunnels, but with only limited

Image 2 The domesticated underground in *Metro 2033*

resources at their disposal, the need to survive is a pressing one. In both titles, players are aligned with Artyom, a young man who has lived his entire life underground. Exhibition, his home station, has been attacked by the Dark Ones, creatures which appeared after the nuclear blast and, in Glukhovsky's words, 'were supposed to be just the opposite of humans … causing people to lose their mind by their mere presence' (2016, n.p.). In an attempt to stop this supernatural threat, Artyom is sent into the wider Metro network in search of assistance and information to help stop this monstrous advance.

During his journey in *Metro 2033*, Artyom tries to uncover the secret of the Dark Ones and, in this attempt, enlists the help of the elite military force of the Rangers, together with whom he manages to uncover the hidden D-6 network (also known as Metro-2), a secondary Metro system build by the Soviet regime and stockpiled with old equipment, resources and nuclear warheads. The game ends with Artyom on top of the iconic Ostankino broadcasting tower, having launched a new missile to destroy an enclave of Dark Ones who have made Moscow their home.[5] *Metro: Last Light* opens where *Metro 2033* ended: Artyom is now a member of the Rangers, having been rewarded for his role in the discovery of D-6, but finds himself in the middle of two opposing sides. The two largest factions within the Metro, the Red stations and their communist ideology on the one hand, and the fascist Reich on the other, are preparing for battle over the tunnels and the resources within them. Focusing on the inhumanity of humans over the supernatural terrors of the first game, *Metro: Last Light* culminates in a last ditch attempt by the Rangers to defend the Metro-2 system. When they are overrun, Artyom is forced to trigger the network's self-destruct sequence, leaving the player with an image of D-6 engulfed in flames—the fate of those inside uncertain.

This narrative arc is compelling in terms of its game design and starting point: the first-person perspective connects players to Artyom, and his story of never having left Exhibition reinforces the players' status as a novice in this world. Both player and character will venture into the Metro for the first time to encounter this new use of the transport system. Underground, the Metro is a place of contrasts, between the dark, unknowable tunnels, and the relative safety of the domesticated spaces of the stations. The latter are steeped in normality, or what passes for it in a post-apocalyptic society: players encounter small cities build within the station spaces, including bars, markets, trading posts and often some version of a local government. Throughout the houses themselves, furniture has been repurposed to mimic rooms and living spaces, often littered with details such as coffee cups, children's toys and washing that has been left out to dry (image 2).

Life has continued since the explosions of 2013: people have found a way to exist, children have been born and some form of order has been established since the need to flee below ground. This sits in stark contrast to the chaos and destruction encountered above ground: soon after leaving Exhibition, Artyom and his companion Bourbon are forced to travel to the surface, to what is described by the latter as 'the dead city', as he bids Artyom a wry welcome home. Looking upon Moscow for the first time, Artyom observes in his journal that he 'felt both fear and sorrow' and 'saw strange beauty in the dark skies and frozen landscape'. However, as soon as the two men are able to return to the tunnels, Artyom notes that the city is 'now a foreign and hostile place'. He is unable to recognize the city and how he lived there, as to him it is now 'an apparition, a nightmare', concluding that 'I have no home but Metro'. Artyom's current experience of the world, and its new order, is made abundantly clear.

In trying to map this new understanding of the cityscape against academic literature on the topic of space, a complex picture emerges, and the work of Augé and Tuan on constructions of space and place will be used to investigate this. In his book *Non-places* (1995), Augé introduces the concept of the non-place, a space of hypermodernity, of a blank spot through which one passes, but never connects with on a deeper level. The anecdote which opens Augé's book cites rental cars, highways and airport terminals as examples of such spaces, and he explains: 'If a place can be defined as relational, historical and concerned with identity, then a space which cannot be defined as relational, or historical, or concerned with identity will be a non-place' (1995, pp. 77–78). The prologue of the work links such non-places to spaces of travel, of commuting, of familiar locations that, to the ordinary person, hold no clear history or deeper meaning; they simply exist.

The same cannot be said, however, for the Metro network, as its symbolism prohibits a similar empty reading of non-place, which, according to Augé, would create 'neither singular identity nor relations; only solitude, and similitude' (1995, p. 103). By contrast, the transport hubs of the real Moscow Metro and its in-game depictions highlight the individual qualities of its stations. The presentation of both aesthetic and ideological content found within the Moscow Metro today, makes it impossible to regard these spaces as having no sense of history or identity. Instead, the construction of this transport system, and its connection to Soviet history and ideals, continue to provide an overlay of meaning that cannot be ignored.

This is further emphasized within Glukhovsky's reinvention of the post-nuclear Metro, in particular in his interpretation of stations as islands, inhabited by different tribes with individual visions, some of which reflect pre-existing political frames. The visits to Reich, Polis and the Red stations in *Metro: Last Light*, in particular, evidence the connection between ideology and design, each station wearing its identity and beliefs on its walls, even sharing a certain iconography (image 3). Indeed, Glukhovsky expresses the centrality of the Russian experience to the franchise, referring to the fall of the Soviet Union as an inspiration for his writing:

> When I was 12 years old, the Soviet Union collapsed. Everything we knew about the history, politics, culture, all our system of values, our entire empire (that seemed eternal, as all empires do) - it all was just CANCELLED by a TV announcement. Overnight. We woke up on the ruins of an empire, on the ruins of our own civilization. (2016, n.p.)

Image 3 A comparison of Artyom's arrival in the fascist station Reich (left) and communist Teatr (right) in *Metro: Last Light*

Later on in his comments, Glukhovsky notes the relationship between his work and the current political situation in Russia: 'I think that Metro 2033 was just an attempt to sum up my thoughts and feelings of our life in Russia in the nineties. Just like Metro 2035 is an attempt to understand where today's Russia is going - and why does it choose to go back into Hell' (2016, n.p.). Interestingly, he compares Russian history to a tragedy, 'a non-ending catastrophe for the last several centuries' and his interest in the idea of how 'we humans never learn from the mistakes of our past' (2016, n.p.). The stations possess an identity and cannot be coded as empty, as non-places. The enduring weight of Moscow's history, and the inevitable repetition of it after the nuclear blast of 2013, as depicted in the novels and games, resonates through its tunnels.

Yet with regards to the post-apocalyptic Metro, there are other aspects that need to be considered. In his definitions of the concepts of space and place, Tuan explains their distinction: 'Place is security, space is freedom: we are attached to one and long for the other' (2014, p. 3). More interesting for the current argument is Tuan's assertion that one can shift into the other, as 'what begins as undifferentiated space becomes place as we get to know it better and endow it with value' (2014, p. 6), and ultimately, 'when space feels thoroughly familiar to us, it has become

Image 4 Some recognizable landmarks remain, such as the Red Square in *Metro: Last Light*

place' (2014, p. 73). The appropriation and domestication of many of the stations point to such a transformation into 'place'. Yet implicit in Tuan's assessments is the possibility for a continuous shift, where there is not just the possibility of space becoming place, but of place becoming defamiliarized and shifting back to abstract, unknowable space. As Augé notes, 'the possibility of non-place is never absent from any place' (1995, p. 107), and any known place is always at risk of losing its status of familiarity and safety. The Metro as imagined by Glukhovsky continuously shifts between this understanding of space, non-place and place, holding these dualities within itself. Artyom's world is made up of underground and overground spaces, perceived as both dead and alive at various times, as stations get overrun and destroyed, and the churches on Moscow's streets become a safe haven for the Rangers. Underground, there is a shift between the stations and the tunnels, between domesticated place and unknowable darkness, yet even within this divide, not all of the Metro feels like home, and this tension between the domestic and the *unheimlich*, between order and chaos and between place and space, frames the way in which players encounter these tunnels, never fully able to know, or trust, what they might find within.

In addition to the visual design of this new world, both titles further illuminate a player's understanding of the Metro universe through chapter introductions and journal entries, written by Artyom, which serve as a form of narration and commentary and, in a way, a historical record. Artyom's journey is current to the player, but in following him, they cannot escape the sense of past and of memory which permeates the Metro. Many of the stations retain their old names, although some have been changed to reflect events that took place after the move underground (with Cursed Station, for example, gaining its name after an attack which left its inhabitants dead). Some reflect, or even mimic, that which lies above ground: a visit to Theatre in *Metro: Last Light* shows that the famous Bolshoi Theatre, located on top of the station, has been moved into the Metro, with a variety of acts performing underground. At the heart of it, however, the representation of the Metro found in 4A Games' titles is a warped version of the existing system, as, through most of the games, the splendour of the existing transportation system cannot be found. The names of certain stations live on, whereas others have been renamed in the aftermath of the bombs, yet the games only show the domesticated spaces which appear to have been constructed behind the outward grandiosity of the Metro, in the rooms and halls that lie beneath the immediate stations and platforms. It is only when Artyom reaches Polis, described as 'the last citadel of culture and civilization', 'a real underground city' which is 'vast, shining, cleaned to perfection', that some of the grandeur of the people's palaces can be seen. The pinnacle of Artyom's journey in the first game, access to Polis opens the door to further palatial depictions of the Reich and Red stations in *Metro: Last Light*, each of which is adorned as the old stations were. This is further emphasized by an understanding of the lineage, with one character commenting on the fact that 'our new world is built out of the debris of the old one'.

At the same time, knowledge of the old world seems to be waning as the overground city and its landmarks are stripped of meaning. The surface locations visited by Artyom in the first game are primarily residential in nature, whereas *Metro: Last Light* takes the young man to the Lenin Library, Red Square and even Lenin's mausoleum, the body of the leader of the revolution now decayed beyond recognition. Yet Artyom, at twenty years old, was only a baby when the bombs fell, largely unaware of the previous Moscow and its symbols. In his journal, Artyom records his visit to Red Square, but only in light of his knowledge of it, based on old images and stories heard from others: 'What is it like, the Red Square? I can't wait to see it. People before used to say it was the heart of our motherland. I'd never been to this place, and saw it only on the old postcards'. He is not alone in this: on their journey, players will encounter many smaller children who have been born after the move to the Metro, who will never have seen the city, or even daylight. Artyom describes them, too, 'The children of Metro... Sometimes I feel that they, who were born underground and who have never seen the Sun, are a whole new species'. The city of Moscow, and with it many icons of Mother Russia, have lost their significance for a generation who have never known these spaces. Russia's history is superseded by the new world order, post-nuclear attack, the overground lost to the landscape of the Metro (image 4).

However, the relationship between the history and the present displayed in the games is more complex than a copying of aesthetics and activities, or even a return to older fascist ideologies. Mighall's interpretation of the urban Gothic refers to the way in which the 'Gothic *of* the city [...] needs a concentration of memories and historical associations' (2007, p. 57; author's emphasis), and, as I have set out, the symbolism with which the Metro system is imbued carry these memories within their very bricks, as do the ruins of the overground, even if they are no longer known by younger generations. Mighall continues that these associations would 'ideally ... be expressed in an extant architectural or topographical heritage, as these areas provide the natural home for ghostly presences of imagined / projected meanings' (2007, p. 57). Thus the ideologies and aesthetics of the real-life Moscow Metro provide a rich tapestry for carrying these meanings, but it also adds complexity to their existence in the games. The digital representation of the stations does not only reflect actual locations but also draws on the urban legends which haunt the Metro (and are cited by Glukhovsky (2016) as one of the influences for writing the *Metro* novels).

An example of this is the D-6 military bunker. The final location of the first game and central to the plot of the second, D-6 is a secret underground system, built during the Soviet era. Inhabiting a strange position between fact and fiction, the D-6 in the games is a reality, a metro within the Metro, stockpiled with weapons and supplies. Despite its ties to the Soviet regime, Artyom's journal raises further questions about the history of these particular tunnels. Instead of solely looking to the labourers of the USSR, Artyom also discusses the efforts of those working for the Tsars, but notes that neither was the first to dig here. The journal describes how he passes through 'even older caves and burrows, built in times immemorial' with those who created them 'leaving only endless corridors going from nowhere to nowhere again behind'. It adds a flavour of the ancient to this feat of engineering, a system of transportation famed for its modernity upon opening, yet which may possess its own, more ancient, secrets: 'Nobody really knows what's there around the Metro stations, above them or below them'. Within the tunnels, old and new, the ghosts of those who passed through still linger. Indeed, both Artyom and his companions encounter shimmers of past events; of children playing without a care on the surface; of an exploding train, killing dozens, when the bombs fell on the city; of his own future, an apartment, a wife and child. The presence of imagined and projected meanings, as described by Mighall, colours the understanding and experience of both the Metro and the surface; its present and its past. Within itself, Moscow and its Metro as represented in the games are marked by their relationship to one another, and how these spaces (including the individual stations) are perceived by the current inhabitants. At the same time, this connection is overlaid by a sense of history, of memories made within the Metro, both before and after the nuclear disaster, and infused into the tunnels themselves. The old and new coexist, overlap and influence one another, and, as the comments and journal entries make clear, leave a profound effect on those who live within the Metro.

Ultimately, then, Glukhovsky's imagining of the Metro is based in spaces of opposition, of under/overground, of home/*unheimlich*, of order/chaos, and in order to map the significance of these binaries, I now turn to Foucault's concept of the heterotopia as a way of understanding this new Moscow. In his essay, Foucault explains how '[certain sites] have the curious property of being in relation with all the other sites, but in such a way as to suspect, neutralize, or invert the set of relations that they happen to designate, mirror, or reflect' (1986, p. 24). He calls these heterotopias, defined as places which are 'are outside of all places, even though it may be possible to indicate their location in reality' (1986, p. 24). More specifically, '[t]he heterotopia is capable of juxtaposing in a single real place several spaces, several sites that are in themselves incompatible' (Foucault 1986, p. 5), which is, in part, linked to the way in which heterotopic sites connect to other spaces. Foucault describes 'a system of opening and closing that both isolates them and makes them penetrable' (1986, p. 26), creating an opposition of access and passability.

On a deeper level, Foucault notes that any attempt to enter them is only an illusion, since 'we think we enter where we are, by the very fact that we enter, excluded' (1986, p. 26). It would be easy to point to the ability to move between stations, and from the underground to the surface, but this exclusion also exists within the experience of the spaces, and the symbolism of the Metro itself. One of Artyom's companions, Khan, simply states that 'I know the tunnel, and the tunnel knows me'. Others speak of the Metro as if it were alive: 'The Metro is a living, breathing thing… with a heartbeat, a soul, - and a mind'. This sits in stark contrast to a trip to the surface by Artyom, during which he describes the city as 'a crypt city. A fallen hero city. A city bereft of its soul'. The famous Russian soul, which sits at the root of any understanding of Russia and its inhabitants, has disappeared. Taken from the capital, the symbol of the nation, it has been buried beneath it, kept alive by those brave souls who are fighting for a better life. The city and its landmarks remain, in some form, but have lost their history and meaning, in favour of the ghosts of the past, and of the people, who have been forced below ground.

The heterotopic nature of the Metro is found in its instability and precarity. Despite being described and coded as domesticated, as ordered and as offering a modicum of safety, the narrative events of *Metro: Last Light* further destabilize this unnatural environment, where the war between Reich and the Reds and the struggle for supremacy between fascism and communism indicates a repetition of history. Furthermore, although *Metro 2033* opens with the line that the Metro is 'our home, the only thing we left', the drive to find a way out continues; the new world order will not remain ordered. A third game in the series was released in 2019, titled *Metro: Exodus*, and from the trailers that developer 4A Games showed prior to its release, players will once more take the role of Artyom, but this time, they are leaving Moscow behind. Using a recently discovered locomotive, still intact after the nuclear attacks, Artyom and his companions set off on a journey across Russia's mainland, ostensibly in search of more survivors. A departure from the relative safety of Moscow's underground spaces, perhaps made uninhabitable after the disasters detailed in *Metro: Last Light*, one can only imagine what dangers still lie ahead. Having been forced underground, and having adjusted to this new world order, there is still a drive to re-emerge, to brave the wastelands beyond Moscow. The safety of staying inside, away from the dead city, is sacrificed for the chaos of the brave new world, yet it is a freedom which may well become more oppressive than the tunnels.

Notes

1. A live action trailer with the title 'Enter the Metro' was released in May 2012 to support the release of *Metro: Last Light* in 2013. An extended version (in Russian) can be found online: https://www.youtube.com/watch?v=cLbj1LMK_ZU&t=293s [Accessed: March 12, 2018].
2. A full overview of the media that form part of the Metro universe (in Russian) can be found here: http://www.metro2033.ru/ [Accessed: March 12, 2018].
3. The connection between Glukhovsky's trilogy and the 4A games is somewhat complex: *Metro 2033* is a direct adaptation of the first novel, whereas *Metro: Last Light* is a sequel to the first game, as opposed to following the plot of the *Metro 2034* novel. Instead, *Last Light* is novelized as *Metro 2035*. However, Glukhovsky has been involved in the writing of the trilogy of novels and both games, linking him closely to the work.
4. *Metro 2033* and *Metro: Last Light* were released again as a bundle, entitled *Metro: Redux* (2014). This version included all downloadable content (DLC) for both games, as well as updates to the graphics. This version was used as the primary source for this chapter. Any quotation without a reference has been taken from in-game text or dialogue.
5. Based on a player's actions during gameplay, they can earn morality points. The number obtained can result in a different ending to *Metro 2033*, in which Artyom decides not to fire the missiles. The series itself considers the destruction of the Dark Ones as canonical, using it as the starting point for *Metro: Last Light*; therefore, I will not discuss the alternative ending here.

References

Augé, Marc. 1995. *Non-places: Introduction to the Anthropology of Supermodernity*. London: Verso.

Bowers, Katherine. 2013. The City Through a Glass, Darkly: Use of the Gothic in Early Russian Realism. *The Modern Language Review* 108 (4): 1237–1253.

———. 2017. Through the Opaque Veil: The Gothic and Death in Russian Realism. In *The Gothic and Death*, ed. Carol Davison. Manchester: Manchester University Press.

Byron, Glennis. 2012. Global Gothic. In *A New Companion to the Gothic*, ed. David Punter. Oxford: Wiley Blackwell.

——— (ed.). 2015. *Globalgothic*. Manchester: Manchester University Press.

Cornwell, Neil (ed.). 1999. *The Gothic-Fantastic in Nineteenth-Century Russian Literature*. Amsterdam: Rodopi.

Foucault, Michel. 1986. Of Other Spaces. *Diacritics* 16 (1): 22–27.

Glukhovsky, Dmitry. 2016. Ask Me Anything. Reddit, December 12. https://www.reddit.com/r/books/comments/5hyvwa/im_dmitry_glukhovsky_the_author_of_metro_2033/.

Jenkins, Henry. 2004. Game Design as Narrative Architecture. In *First Person: New Media as Story, Performance, and Game*, ed. Noah Wardrip-Fruin and Pat Harrigan. Cambridge, MA: MIT Press.

Kuznetsov, Sergey. 2016. The Symbol of the Dynamic Capital: A New Image for the Moscow Metro. In *Hidden Urbanism: Architecture and Design of the Moscow Metro 1935–2015*, Sergey Kuznetsov, Alexander Zmeul, and Erken Kagarov. Berlin: DOM Publishers.

Luckhurst, Roger. 2002. The Contemporary London Gothic and the Limits of the "Spectral Turn". *Textual Practice* 16 (3): 527–546.

Meuser, Philipp, and Anna Martovitskaya. 2016. Infrastructure of a Hidden Urbanism: Introduction. In *Hidden Urbanism: Architecture and Design of the Moscow Metro 1935–2015*, Sergey Kuznetsov, Alexander Zmeul, and Erken Kagarov. Berlin: DOM Publishers.

Mighall, Robert. 1999. *A Geography of Victorian Gothic Fiction: Mapping History's Nightmares*. Oxford: Oxford University Press.

———. 2007. Gothic Cities. In *The Routledge Companion to Gothic*, ed. Catherine Spooner and Emma McEvoy. New York: Routledge.

Phillips, Lawrence, and Anne Witchard (eds.). 2010. *London Gothic: Place, Space and the Gothic Imagination*. London: Continuum.

Totten, Christopher W. 2014. *An Architectural Approach to Level Design*. Boca Raton, FL: CRC Press.

Tuan, Yi-Fu. 2014. *Space and Place: The Perspective of Experience*. Minneapolis: University of Minnesota Press.

Warwick, Alexandra. 1998. Urban Gothic. In *The Handbook to Gothic Literature*, ed. Marie Mulvey Roberts. Basingstoke: Macmillan.

Smith, Harvey, and Matthias Worch. 2010. What Happened Here? Environmental Storytelling. Presentation at 2010 Game Developer Conference. https://www.gdcvault.com/play/1012647/What-Happened-Here-Environmental.

Zmeul, Alexander. 2016. Excavating Underground Archives: Architectural History of the Moscow Metro. In *Hidden Urbanism: Architecture and Design of the Moscow Metro 1935–2015*, Sergey Kuznetsov, Alexander Zmeul, and Erken Kagarov. Berlin: DOM Publishers.

'Everything Is True': Urban Gothic Meets the Chthulucene in Multiplayer Online Game, *The Secret World*

Tanya Krzywinska

Gothic appears in games of all stripes, appearing across genres and platforms. Alongside Science Fiction, Gothic is one of the most diversely deployed generic languages found in games and often the two appear in the same game. In *Space Invaders* (1978, Taito), wave after endless wave of those pesky pixels demanded quick reactions and deep focus from the player, creating inklings of infinite space and a metronomic relentlessness that countered the rhythms of more human-centred narratives. The inky blackness of the *Space Invaders* backdrop conjured up the infinite sublime, a place where no one can hear you scream as you succumb to the vertigo of the game's inexorable downward march. In early video games the black screen of space was the effect of limited resources in terms of graphics and computing power. These constraints produced some interesting creative design solutions, as with the use of video footage in games like *Phantasmagoria* (1995, Sierra) or abstraction and synecdoche in games such as *Haunted House* (1982, Atari) and *Clock Tower* (1995). While such constraints afforded creative solutions and media-distinctive iconography, it was only with the introduction of

---

T. Krzywinska (✉)
Falmouth University, Falmouth, UK

H.-G. Millette and R. Heholt (eds.),
The New Urban Gothic, Palgrave Gothic,
https://doi.org/10.1007/978-3-030-43777-0_8

techniques that allowed 3D space to be recreated in games, alongside the increasing capacity for graphical fidelity, that conditions for a more realistic simulation of the urban in games became possible. As we will see the Urban Gothic of contemporary games draws together all these techniques.

Gothic has roots in earthly soils; coffins rot in an overgrown graveyard, a figure dances widdershins on All Hallows Eve, in the distance glides the ghost of a weeping nun. Many games have sought to remediate nodal Gothic texts such as Mary Shelley's *Frankenstein; or, The Modern Prometheus* (1823) and Bram Stoker's *Dracula* (1897) and Horace Walpole's *The Castle of Otranto: A Gothic Story* (1765); the latter provides a blueprint for the typical horror dungeon-crawler. Literary adaptations in early video games were common ways of calling on Gothic, often relying on written text as a principal mode of creating ludic and narrative engagement. More cinematic modes of representation had to wait for the development of resources that allowed for greater visual verisimilitude. Later games deployed 3D space (within which players gain a sense of moving through a topos), alongside first- or third-person perspectives and varying degrees graphical realism. Not only did this make the Rural and Folk Gothic possible in a simulative sense, as with *Clive Barker's Undying* (2001, EA) and *Everybody's Gone to the Rapture* (2015, The Chinese Room), but it also meant that more complex urban spaces could be realised and designed to support a variety of gameplay types.

In games that deploy Urban Gothic the soil does not pulse to its own othered rhythms as it does in the pastoral Gothic. Instead urban dirt is made up of human detritus, sooty and morally loaded; even out in the suburbs, where the pastoral mixes with the urban, human soil erupts in neatly tended gardens. In games, the environment is never incidental: it provides story and importantly shapes the ludic experience. This essay considers the patterns and meanings of the Urban Gothic in games. It begins by teasing out certain themes and tropes that colour this ludic souterrain, paying attention to developments in games technology that have made possible a convincing simulative rendition of the Urban Gothic. This is exemplified by multiplayer online game *The Secret World* (2012, Funcom), relaunched as *Secret World Legends* in 2017, a game that uses urban myth and conspiracy as a means of articulating urban Gothic and discussed in some detail later. For now, I'll continue to draw out some of the significant features of the contexts within which the urban appears in Gothic games before moving on to discuss core vocabularies.

Urban Gothic developed in nineteenth-century fiction by, for example, Charles Dicken's *Bleak House* (1853), Robert Louis Stevenson's *Strange Case of Dr Jekyll and Mr Hyde* (1886) and Bram Stoker's *Dracula*, and now provides the basis for modern games. Urban Gothic games regularly seek to present cities, towns, conurbations and the suburbs as full of threat, darkness and the degeneration of humanity. Very often Urban Gothic games anthropomorphise the city or town itself, making it a character, often demonic and best exemplified by the eponymous town of early Silent Hill games (1999, 2001, 2003, Konami). This is a trope that has its roots in Film Noir.

In addition to the city as threat, there is also a certain strain of Urban Gothic games whose lineage is earthed in vampire fiction. Urban vampires generally can be traced back to des Esseintes, the urban aesthete of JK Huysmans' Decadent novel *À rebours* (1884) as well as to Anne Rice's city-dwelling vampires. These figures have their ludic counterparts in White Wolf's Vampire: The Masquerade franchise. Alongside its original table-top game first released in 1991, where players imagination is integral to engagement, the Masquerade world also includes video games, *Redemption* (2000, Nihilistic), *Bloodlines* (2004, Troika) and *Bloodlines 2* (due for release 2020, Paradox). All these games allow players to create an urban vampire character choosing various traits and physical features. Players are tasked with surviving in a hostile urban environment, remaining hidden from the police yet feeding off unlucky humans. While players call on their imagination when playing the table-top role-playing scenarios; the video game versions interpret and realise character, location and action. Over the various video game releases in the series, animation, visual complexity and the affordances of space have been increasingly more expansive and simulative. This lends the urban settings of the games greater weight and meaning. *Bloodlines 2* is currently selling itself on realistically rendered characters in a vast, simulated Seattle cityscape. In noir-ish vein, there is even a nod to the city itself as vampire in the trailer, 'this city will bleed you dry'. This latest incarnation is cinematic in its visual qualities and the ratio format reflects that; the visual, spatial and auditory detail of the Urban Gothic evoked in literature and displayed photographically in cinema has become more literally realisable as game technologies have become more sophisticated.

Such detailed environments have also become more integral to gameplay and in some cases are able to better realise the themes that are integral to the Urban Gothic. In their high-resolution urban world,

Masquerade's player-character vampires must navigate vampire as well as human laws. It is this that forms the design of the player's ludic experience as well as the design of the game's urban environment. There is therefore a thematically laden morality system built into the games: each vampire has 'humanity points' that are lost when killing innocents, so possible behaviour must be managed, and the urban spaces warily navigated. It is here that the social urban context created by the game also becomes instrumental to gameplay design, rather than solely providing an atmospheric backdrop.

Gothic games that deploy urban settings frequently call on aspects of 'noir' and I will now focus on how noir is used in games. Noir is by far the most commonly used vocabulary within Urban Gothic games Generally, in noir the urban is consonant with murder. Frank Wedekind's play *Pandora's Box* (1904) and Fritz Lang's '*M*' (1931), both of which reference the Whitechapel murders of the 1880s, provide early examples. Death is inevitably at the heart of things, linking crime fiction to Gothic. As Borde and Chaumeton have said in their essay 'Towards a Definition of Film Noir' written in 1955, 'In every sense of the word a noir film is a film of death' (2002, p. 19). They also identify key noir-ish features, features that have become staples within Urban Gothic games. These include: realistic settings; moral ambiguity; a passive rather than heroic protagonist; plot twists; uncertainty of motivations; oneirism and weirdness. To this list we should also add pessimism, following Packer and Stoneman's focus on pessimistic rhetoric in popular culture (2018, pp. 20–24).

Oddly, what Borde and Chaumeton don't mention is the way in which the city or the urban provides the condition for Gothic to be rendered in noir style. In the context of games, the urban also provides geometry, and, as with architecture generally, it is from geometry—the basis for 3D modelling tools—that game environments are built. Game spaces are far more easily built and designed in a geometric context rather than in the organic melange of the rural. Urban spaces are full of geometry, architectural lines and symmetries that are far easier to manage in a game engine. Geometric structures and *chiaroscuro* lighting design bring drama and demonic life to an urban location that are then easily co-opted for Gothic intent. With its clean, jagged lines Deco-style noir brokers no squishy, home comforts or gauzy mysticism. In games, solid architectural forms become part of the ludic design. As urban spaces are mappable systematised environments, street names and signs navigate players purposefully through vast complex sand-box spaces while providing back streets and dives to secrete noirs' demi-monde.

As with Film Noir, and as the Masquerade example above indicates, the Urban Gothic trades on anxiety, paranoia and conspiracy. Such noir-ish thematics provide the context and atmosphere required for those Gothic games that seek to play cat and mouse with a player's sense of mastery over events. Such games within games ratchet up a sense of panic, jeopardy and uncertainty for players. Masquerade's vampiric rule set, for example, means that players are not cast as individualistic, free-agent superheroes, instead they are vulnerable and need to exercise caution to survive in a complex social world. These vampires know what it is to feel fear and act cautiously. The urban and social context puts obstacles in their way; these vampires are not therefore Lords of their domain as Dracula was in his rural homeland (although we should note that it was the complexities of urban undead living that bested Dracula in Stoker's narrative).

Noir-ish elements are found across a vast range of game genres and platforms. Some games focus more on the gangster and criminal caste of noir, as appears in the Grand Theft Auto (henceforth, GTA) or LA Noire series—Gothic can however seem more distant within such games where magic or the occult do not feature. The Max Payne series for example is a fast-paced first-person shooter set within the urban criminal underground of New York with no presence of the supernatural or magic. Max's existential and pessimistic outlook, episodes of altered consciousness and the use of names derived from Norse myth tailored to a contemporary urban context, make for a darker experience than that of GTA and LA Noire, thereby the Max Payne series has a stronger claim to the Urban Gothic and new weird.

More squarely placed in Urban Gothic soil, if by way of a fantasy rather than referential world, is the adventure/conversation game *Discworld Noir* (1999, Perfect/Teeny Weeny,) based on the world created by Terry Pratchett. The game creates its blend of noir using ingredients derived from Dashiel Hammett's *The Maltese Falcon* (1930), H. P. Lovecraft's Cthulhu mythos mixed with a jigger of genetic occultism. *Discworld Noir* relies heavily on on-screen written narration and instructions to tell its story and its environments lack the visual simulative density of more recent games. This represents a very different way of immersing and locating the player than is found in fully rendered 3D games, where environmental storytelling is fully in play. In a similar vein to *Discworld*

Noir, *The Wolf Among Us* (2013–2014, Telltale), another more recent adventure game, based on the comic series Fables (2002–2015, Vertigo) written by Bill Willingham, is set in contemporary New York. Having fled Europe to become refugees in the city, characters from European fairy tales must find the murderer in their midst. Placing rural Gothic characters and stories in an urban context throws up some interesting dramatic juxtapositions. The urban setting of this game provides the context for a noir-ish faceless murderer tale that underscores the characters' anxiety around their need to normalise to their 'new' world. With less gameplay and an 'on-rails' linear narrative, the game takes a more literary, figurative approach to its version of Urban Gothic. In this, it is in line with *Discworld*; both are best described as interactive fiction than as games where game play verbs take precedence.

There has also been some transmedial traffic from video games into other noir-ish and new weird fiction, thereby altering and expanding Urban Gothic tropes. The multilayered, opaque cities of China Miéville's fictions, i.e. *The City and The City* (2009) and *Un Lund Dun* (2007) are indebted to games as it is in games that single spaces often have many layers; Miéville's role as a game writer may well have originated the way he figures space in these novels. This formal strategy appeared in games as a means of managing available computing resources. Giving 3D assets new 'skins' (textures) meant that one space could be used in two or three different ways. This distinctive aesthetic chimes with that of the underground, the hidden souterrain, and decadent demi-monde as figured by Alexander Dumas (1855). The layering of space in Urban Gothic allows for an expression of Alt.spaces, the disavowed, the transgressive and the unconscionable. Early games within the Silent Hill series provide a representative example of this aesthetic spatial economy—areas of the eponymous town have 'hellish' *and* 'normal' incarnations: same space in the game and the same architecture, but dressed in different textural digital skins. In games this is a resource-led textual strategy established earlier by *Legacy of Kain: Soul Reaver* (1999, Crystal Dynamics), where the player-character must switch perceptual modes between material and spectral planes to solve puzzles. This doubling invests a given location with different signifying and ludic elements. The split city mode then carries the potential to represent shadow worlds, rendering space liminal and uncertain, as it occurs in Miéville's *Un Lund Dun*. In noir, the urban is treacherous, ambiguous and has a mind of its own; no longer a domain that is in the service of 'man', it has become unpredictable and othered.

In games or in any fiction, setting and its mise-en-scene are always tightly controlled by authors. The noir urban space must function to signify, to tell a story environmentally through its characterisation, and, in games, it must further be architected to support gameplay activities. The Assassin's Creed series has set the industry benchmark for creating believable urban spaces in games. The creation of increased depth of field allowing expansive city vistas rendered as crisp high-resolution images that are incrementally redrawn as the player traverses the space has set an expectation about the rendition of urban spaces in contemporary games. With Assassin's Creed, the stage was set for Gothic games to create complex urban spaces that are capable of providing the type of sublime visual expansiveness found in Pieter Bruegel's painting *Hunters in the Snow* (1565, Oil on Wood).[1]

Urban contexts provide game designers with a range of possibilities with which to engage players in Gothic gameplay. Providing social, psychological and spatial mechanics, as we have seen with *Vampire: The Masquerade* series and other examples mentioned above. Urban space is never therefore simply backdrop. Instead it has substance, is often an active protagonist and designed around core gameplay activities. Whether it is a fantasy urban space (Discworld), a real space (Los Angeles or Seattle in Vampire: The Masquerade and, London or Seoul in *The Secret World*), realised in a figurative or simulative way (*Wolf Among Us* or *Bloodlines 2*), or experienced as linear or free-roaming, the same principle follows; backdrop is never simply setting. What must be present in that depiction of urban space for a game to earn the title Urban Gothic is an existential pessimistic edginess, where monstrosity breaks the bounds of self and other—of space and time—and following that, no substantial sense that the player-character is a straightforward hero with mastery over the space. While other approaches are effective, abstract representation is no longer the only choice for game developers. With greater computational resources in play, complex visual landscapes are now easier to deploy, making it easier to realise the nineteenth-century Gothic trope where cities are places of disease, crime, transgression and exploitation. Through technologies such as high-resolution graphics, motion capture, fluid dynamics and AI, video games now have the ability to render and animate believable humans and creatures and create complex traversable spaces. Such tech-driven shifts challenge Leonard Cassuto's view that

Urban Gothic is the domain of human rather than supernatural monsters (2017). I submit *The Secret World's* extensive and varied demonology to support the point and to provide a closer look at the realisation of Urban Gothic in the context of a multiplayer online game.

Urban Gothic, Cththulucene and *The Secret World*

As is common with so many examples of recent popular culture, *The Secret World* creates a world—taking advantage of the simulative aspects of game technology, it is a world set in the contemporary 'real' world including multiple recognisable urban spaces. This game also provides social spaces, or what Ducheneaut and Moore (2004) call 'third space' (home and work being the first and second), that operate much as a bar might in the physical world. As Dimitri Williams (2006) notes, multiplayer online games are designed around social interaction between players, and, further, to create the conditions for what Lisabeth Klastrup calls 'emergent player stories' (2009) which are likely to arise from group's working to overcome gameplay challenges. The advantage of creating a world designed as a social space is that it can support a vast range of elements and is infinitely expandable, perfectly suited for franchises and multiplayer online games. Just as Tolkien set out to build Middle-earth and its ecologies as a toolkit for others to use, *The Secret World* has within it many different types of gameplay activities, different diegetic cultures, a multitude of micro and macro storylines as well as supporting role-playing, emergent behaviours and fan culture. While a world has inevitably different types of spaces within it, this game has a high concentration of urban ones. It has cities that we recognise, such as London, Tokyo, Seoul and New York, plus a host of towns of various sizes and displaying different types of cultures dotted across the game's world. There are also many rural locations, some coastal, others arboreal and desert. These sparsely occupied spaces make the urban spaces more solidly 'urban' because of their differential juxtaposition. Equally, the rural locations benefit in terms of their ability to be experienced as places of solitude (cities can be places that are busy with other players doing daily tasks, hanging out with friends and managing their resources).

The Secret World's urban spaces employ a wide range of vernacular and stylistic vocabularies. While the whole world communicates human imperilment and moral ambiguity, its noir-ish characteristics coexist with other Gothic modes: part of the game is set in Egypt therefore co-opting what Roger Luckhurst calls 'Egyptian Gothic' which has a strong occult flavour (2012) and the myth of the mummy's curse (Frayling, 1992). Another area nominates Vampire Gothic in the part of the game set in Transylvania. The gameplay proper opens on Solomon Island, off the coast of Maine and draws on American Gothic staples. The first of which is the Cthulhu mythos, a fictional world referred to by several different writers that is populated by monsters originating from stories written by HP Lovecraft (as distinct from Haraway's differently spelled Cthtulhucene, but sharing a relationship to the tentacular. See Image 1 for Lovecraft's drawing of the monster Cthulhu and its characteristic tentacles). The second is the literature of Stephen King (King is represented as a character, a horror writer living in a lighthouse).

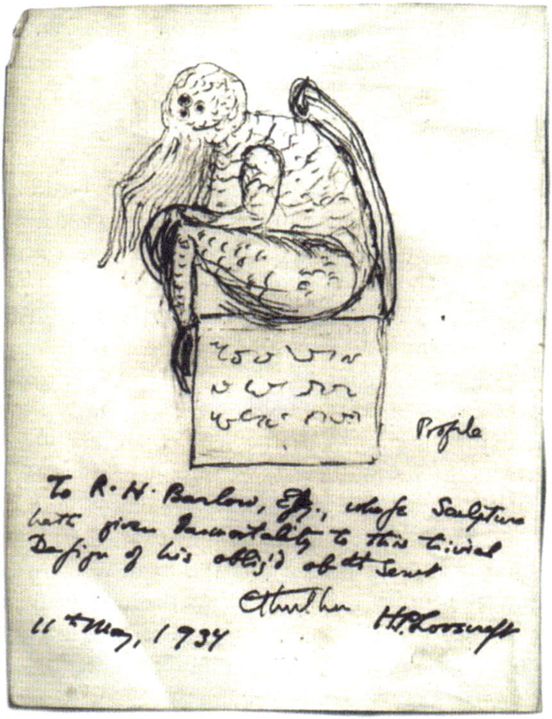

Image 1 Lovecraft's sketch of the Great Old One, Cthulhu

However, what unites its distinctive regional and stylistic vocabularies is the game's creation of a sense of paranoia. Noir often links paranoia with the urban as if it belongs best in busy social space where conspiracy is most able to thrive. *The Secret World* is very much indebted to the occult conspiracy paranoia-inducing scenario set out in *The Illuminatus! Trilogy* (1975), novels written by Robert Anton Wilson and Robert Shea (*The Eye in the Pyramid, The Golden Apple* and *Leviathan*), Wilson's subsequent essay-fiction Cosmic Trigger series (1977) and the card-based spin-out game *Illuminati: The Game of Conspiracy* (1982) designed and published by Steve Jackson games.

Echoing the language of the *Illuminatus! Trilogy*, the game heralds the player with the maxim that 'Everything is True'. This is a part quote from Albert Camus' existentialist novel *L'Étranger* (1942)—'Everything is True, Nothing is True' is the complete phrase. In the context of the game, the aphorism refers specifically to conspiracy theories, myths and urban legends. In effect, *Everything is True* radically challenges the notion of absolute truth, rendering meaning infinitely ambiguous. As a literary effect, this places all 'into the abyss' (*mise en abîme*). A similar aphorism is used in the Assassin's Creed series, here 'Nothing is true, everything is permitted' (this is the eponymous creed)—here it provides a defiant statement of amoral free will; this game cannot therefore be claimed for the weird Gothic where entropy and the inability to act holds court. By contrast, *The Secret World*'s usage of *Everything is True* means that there is no longer any discernment possible between truth and fiction; we've descended into a semiotic hall of mirrors typified and noir-ished as a case of weird, delusional paranoia. This suits the urban myth and conspiracy theme of the game. Paranoia is then the rubric of the entire game where all monsters and urban myths of every type are now potentially real. Taking the 'noir' urban climate to the regional diversity of the game's world, it is thematic *and* semiotic. The requirement that players must use the 'real' internet to help solve certain puzzles blurs the diegetic parameters of the game. Additional blurring comes from the information that the game developers have planted across the 'real' web. In breaking the fourth wall, the game's 'magic circle' and its diegesis, is drawn far wider than is usually the case. Given that the world wide web is infested with conspiracy theories, false news, myths and misinterpretation, its co-option by the game underscores its ability to implicate the player in its fictional web of twenty-first-century paranoia. As part of this matrix, it is clear in the game that technology will not provide a solution to the ecological and existential threat posed to humanity, so the game builds its aesthetics of paranoia around aspects of the Anthropocene.

To help explore and substantiate this idea, it may help to define paranoia in a psychological context. Charles Rycroft defines it thus, 'Functional psychosis characterised by delusions of grandeur and persecution, but without intellectual deterioration. In classic cases of paranoia, the delusions are organised into a coherent, internally consistent delusional system on which the patient is prepared to act' (1968, p. 125). In other words, discernment between fantasy and reality is subject to failure, demonstrating that our grasp on reality is fragile and even where we are 'sane' our internal world colours our interpretation and indeed perception of what is thought to be external. As Mark Fisher notes, 'The weird thing is not wrong, after all; it is our conception that must be inadequate' (2016, p. 15). To which we might add that the ability to read and enjoy fiction itself is based in our facility to temporarily delude ourselves. Configurations of *mise en abîme* are the stuff of paranoia on which the house of weird Gothic is built. Other games, *Eternal Darkness* (2002, Silicon Knights) and *Call of Cthulhu Dark Corners of the Earth* (2005, Bethesda) draw from the same weird pool and as Bernard Perron notes, player-characters of these games are 'affected by various distortions' (2018, p. 275). Uses of *mise en abîme* manifest in Urban Gothic contexts where meaning has become critically unstable and the geometries have turned monstrous. *The Secret World* helps to see how this paranoid schema plays out in the game and in the particular Anthropocentric caste of the urban spaces it creates.

The player-character awakes into the game world in a modern, functional apartment. We see our character pushing balls of energy around, a power newly acquired. Before this power becomes normalised as a superpower, they are approached by the faction that rules the city that they live in. Prior to entry into the game proper, optional videos provide information about the three occult factions from which the player must choose—it's clear that none are fully trustworthy and each regard you as a pawn. The Templars operate like an empire, with rules and traditions. The Illuminati have 'stocks, hell and compromising pictures of angels' plus a long history of manipulating business and governments for their own gain. The Dragons seek to see and manipulate patterns in chaotic systems and use them to bend reality. This game world is not then a replication of the hero's journey: a spiralling process of dissolution and dissemblement unfolds. The player-character is then informed visually in a dream that they must choose, 'lose all or become a god – the choice is yours'. Of course, there is no choice and whatever faction you have chosen the outcome is the same: you've been pressganged by occulted forces that you neither know or understand. And, no matter which faction you choose, they each have you under constant surveillance. In effect, and even before hitting the end of the prologue, the usual definitions of fantasy and reality, choice and determination, are put into question. It becomes clear retrospectively that the little, apparently incidental, bee that flew in through the apartment's window and into the sleeping player-character's mouth in the opening cut-scene of the game was in fact what chose us, heralding the acquisition of a power we did not choose, over which we have little control. This little bee seems out of place in the city—a benign insect that tells another story: a case of the bees telling us? Something is clearly going on; the tidy apartment is now in disarray, uncontrolled power arcs through the player-character's body. The usual rules no longer apply. Even the newly established player-character identity is quickly undermined as the player soon finds themselves in another character's skin, experiencing an event that occurred in the past—time, space and self are no longer stable. Waking again, this time the player's bedroom walls give way to a graveyard, that classic Gothic location, a suitable place for the player to learn the basics of the game's interface and where the dead are infinitely expendable. In terms of gameplay interface, the design of the input controls are not standard and therefore not easily mastered. This makes play somewhat awkward, requiring some contorted manoeuvres at the outset. Nothing and Everything is True.

It is clear from our starting zones that we are urban dwellers; 'cool' contemporary characters models are available for the player to choose. There is clearly an urban concern for clothing; in each hub city there are clothing stores where players can tailor their fashion encouraging personal expression and backstory creation; even in the face of cosmic horror we should adopt sharp contemporary apparel. No cloaks or chain mail here. The urban spaces in the first section of the game are economically drawn, but what is small in scale in terms of explorable space is compensated by detailed architecture, giving a semblance reality through depth. There is also a whole world awaiting exploration and providing a host of myths urban and otherwise to encounter. *The Secret World's* urban space does however have an exotic, weird caste. London is full of street-based occult references, evident in the names of the streets, shops and pubs there (e.g. Saintsditch, Lud Ale Brewery, The Horned God) and the templars control the police force who have the city on lockdown, justifying the presence of barriers.

New York houses some crazy illuminati body-jackers intent on pumping the player-character full of psychedelic drugs located in a car park and the entire zone has a more hectic feel. This is the home of the all-seeing eye, the Illuminati. Seoul is calmer, cleaner, marrying traditional Korean architecture with electronics stores. We learn early on that both Tokyo and Times Square have been attacked by a variety of cosmic horrors. All these places connect the player with their own contemporary urban world—a place not heretofore the subject of multiplayer games, which tend to be set in fantasy worlds as is the case with *Everquest* (1999–, Verant/Sony), *Elder Scrolls* (2014, Zenimax/Bethesda) and *World of Warcraft* (2004, Blizzard). The deployment of these real urban spaces in the game further works to demonstrate that the fabric of human reality is at stake, about to be swept aside by unfathomable cosmic forces that only occult knowledge and perhaps the bees can help to combat (Image 2).

The first assignment that the player-character is tasked with is on Solomon Island, off the coast of Maine, US. Monsters have come in from the sea and the dead have awakened, hungry for living flesh,

Image 2 Times Square under attack in *The Secret World* prologue

another Gothic trope and one that the game often connects to—environmental exploitation. While running around the streets of Kingsmouth (note the reference to both Stephen King and Lovecraft's fictional town 'Innsmouth'), we are enmeshed in a world of urban myths, conspiracies—old and new—and occult histories. Out at the Island's airport, Black Ops activities are taking place and the player encounters a pair who allude to the *X-Files*' Mulder and Scully; both perplexed by events and demonstrating that liberal referential use of conspiratorial fiction in the game. Kingsmouth itself is permanently dressed up in Halloween garb, shrouded in a lurid orange haze out of which legions of the undead and sea monsters emerge. Nordic warriors and Draugrs (Old Norse for 'revenants') are encountered along the harbour and beach-front; they can be found in the town's woods overseeing the undead working to unearth the bodies of deceased townsfolk and those killed in witch-hunts and industrial disasters. It is unclear what their purpose is, bar killing the living—something more than that is definitely afoot, as indicated by the ravens that lead the player to sites of occult rituals inhabited by powerful revenants.

Quests vary from the staple killing of monsters, through to lore-gathering designed for players less focused on ratcheting up their skills and more involvement in the deep storylines that the game contains. Most innovative of these are puzzle quests that involve the player in seeking out information or solving occult riddles. In these the player must look outside the game often to relevant literature, i.e. Poe's story 'The Gold Bug' or into occult history—for example the quest 'Digging Deeper' requires players to decode masonic sigils to unlock a tomb (for more see Krzywinska, 2014). These quests are most strongly aligned to the conspiracy and paranoia theme of the game and, in the case of Solomon Island, all its urban spaces are deeply embedded in occult machinations, having been built and designed by members of the Illuminati. Evidence of this urban occult authorship is littered around the entire island, are conspicuous in the quests like Digging Deeper, and involve a high degree of environmental storytelling is in place to provide the game's lore. The island's lighthouse provides an example of a functional building that is also emblematic of the all-seeing eye of the Illuminati whose symbol is the eye in the pyramid and it is also the residence of Sam Kreig, writer of horror fiction. As with the other factions, surveillance is core to the power over the diegetic world and over you, the player-character and this underwrites the aesthetics of paranoia at work in the game.

Urban myths are closely bound into larger conspiracies and occult factions within the game; often in a spiralling set of complex intertextual connections and references which in a paranoid mindset are given fallacious credence. For example, *Black House* (2001), a novel by Stephen King, who regularly uses Maine as a setting for his fiction, features children that are being killed and cannibalised by a character known as The Fisherman who is, in turn, an agent of the Crimson King (a take on the King in Yellow of Robert W. Chamber's eponymous novel [1895], and whose 'yellow king' guise is adopted by the cult-loving back-woods psycho killer in the first True Detective series [2004]). *The Secret World* also has its version of the Black House and its dedicated questline. The house stands at the edge of the town, is suitably coloured and emits strange noises; it is also surrounded by a powerful energy field that resists entry—players must find a secreted way into the house. Earlier the player is told that a woman was burned alive inside the house by townsfolk and you are tasked with putting her soul to rest. Once in, the house throws debris at you—it's clearly alive, possessed and is invested with all those many urban myths around the 'terrible place', a derelict, unkempt house. There is a clear sense of urban decay here as part of the anthropic landscape—a place where questionable human actions transform into entropic imagery. To quell the concomitant anxiety and attempt to restore (the) 'home', the player must perform rituals to cleanse the house including scattering the ashes of the burned woman in the Miskatonic river. The opening quest text tells us that 'The burnt-out shell of the Black House sits like an accusation among Solomon Island's picket fences and townhouses'. As is so apparent with the urban myths that are common in small-town horror, cities are not the only places to harbour monstrosities and paranoia. In addition to meta-texts and urban myths embedded in the Black House narrative, completion of its quests produces different faction responses that show clearly how they are manipulating you; undermining in different ways any sense of heroic agency. The site of inaction is that of the Gothic and where the conservation of physical energy generates psychic energy in the form of anxiety (Image 3).

The Secret World's people and places are filtered through numerous intertextual references that work to thicken the game's spiralling, metafictional narrative and thereby challenge the assumed boundaries between reality and fiction. The result is a psychoactive blend of urban myth and Gothic, the aim of which is to create a strong sense of the vertiginous nature of the weird for the player that has hitherto not been

Image 3 The Black House that stands at the edge of town in *The Secret World*

achieved within populist games. Players may gain some physical mastery over the game's interface but in the context of the weird and urban, where factions, histories and interpretations constantly slide, there is an escalating sense of uncertainty and paranoia that makes for a weird and Gothic contemporaneous means of overturning the usual certainties and predictabilities found in video games. The game provides an exemplar of the ways in which the Urban Gothic has (optimally) been deployed in the context of video games.

The Secret World's monsters are in different ways connected to injustices of the past, a Freudian 'return of the repressed' spewed up to take revenge on the living. Some of these are related to the treatment not just of other humans but of the environment. The game refers to Anthropocene in this Gothic way, and also connects with Donna Haraway's alternative conception of the epoch, the Chthulucene, which she uses as an alternative to the 'man-centric', marginalising and technocratic uses of the Anthropocene, 'We need another figure, a thousand names of something else, to erupt out of the Anthropocene into another, big-enough story' (2016, n.p.). We hear another story in the buzzing of the game's enigmatic bees and that little bee that afforded the player-character power at the start of the game. These bees seem lead to an elsewhere and a very different story than appears on the Anthropocentric surface of the game, where Lovecraft's Cthuhulean monsters represent a deep pessimism about the fate of humanity and the world. The bees are also allied to the many mentions of Gaia in the game, as Haraway says, 'Gaia was and is a powerful intrusive force, in no one's pocket, no one's hope for salvation, capable of provoking the late twentieth century's best autopoietic complex systems thinking that led to recognizing the devastation caused by anthropogenic processes of the last few centuries, a necessary counter to the Euclidean figures and stories of Man' (2016, n.p.). *The Secret World*'s occult focus enables the game to look beyond Euclidean geometry, bureaucracy and other human constructs and while players might run around looking for power, the game refers to an othered Chthulucene power located within the realm of the bees. As Haraway tells us, 'In many incarnations around the world, the winged bee goddesses are very old, and they are much needed now' (2016, n.p.).

As we have seen, Urban Gothic is represented in games in ways that are increasingly more responsive in terms of activities for players as well as more visually realistic and simulative. With virtual reality, mixed reality and augmented reality platforms increasingly available for the domestic market, the possibility of creating extremely effective immersive perceptual spaces is now upon us. Horror works extremely well in these immersive contexts. Augmented Reality games such *RoboRaid* (2018, Microsoft), overlays and maps images of interstellar robots such that they appear to come out of the walls in the room you are standing in. Other AR games, such as *Knock of the Dead* 'Your own home compromised', *Holophobia: Spiders* (which is meant to be therapy!) or *Night Terrors: Bloody Mary* (2018, Imprezario) for smartphones and *Dreadhalls* (2016, White Door), *Alien Isolation: MotherVR Mod* (2017, Creative Assembly/Nibre), *Kobold* (2018, Another World) *The Exorcist: Legion* for VR, show that there is now the capability to create games that have an even greater capacity to generate paranoia, anxiety and erase the frameworks that signify 'fantasy'. I eagerly await an Augmented Reality version of the likes of *The Secret World* that mobilise potent and rich aesthetics of paranoia and conspiracy within the transformative rubric of Haraway's Chthulucene where the Urban Gothic reminds as that 'human beings are with and of the earth' (2016).

Note

1. Got to https://artsandculture.google.com/asset/hunters-in-the-snow-winter to view a high-resolution image of the painting.

References

Borde, R., and E. Chaumeton. 2002. *A Panorama of American Film Noir*. San Francisco: City Light Books.

Cassuto, L. 2017. Urban Gothic. In *The Cambridge Companion to American Gothic*, ed. J.A. Weinstock, 13–82. Cambridge: Cambridge University Press.

Ducheneaut, N., and R.J. Moore. 2004. The Social Side of Gaming: A Study of Interaction Patterns in a Massive Multiplayer Online Game. CSCW '04: Proceedings of the 2004 ACM conference on Computer supported cooperative work, New York. Available at: http://dl.acm.org/citation.cfm?id=1031667.

Fisher, M. 2016. *The Weird and the Eerie*. London: Repeater.

Frayling, C. 1992. *The Face of Tutankhamen*. London and Boston: Faber and Faber.

Haraway, D. 2016. Tentacular Thinking: Anthropocene, Capitalocene, Chthulucene. *E-Flux Journal* 75. https://www.e-flux.com/journal/75/67125/tentacular-thinking-anthropocene-capitalocene-chthulucene/.

Klastrup, L. 2009. The Worldness of Everquest: Exploring a 21st Century Fiction. *Game Studies* 9(1). http://gamestudies.org/0901/articles/klastrup.

Krzywinska, T. 2014. Digital Games and the American Gothic: Investigating Gothic Game. In *A Companion to American Gothic*, ed. C. Crow, 503–515. Oxford: Wiley Blackwell.

Luckhurst, R. 2012. *The Mummy's Curse: The True History of a Dark Fantasy*. Oxford: Oxford University Press.

Packer, J., and E. Stoneman. 2018. *A Feeling of Wrongness: Pessimistic Rhetoric on the Fringes of Popular Culture*. University Park, PA: Pennsylvania State University Press.

Perron, B. 2018. *The World of Scary Video Games*. New York: Bloomsbury.

Rycroft, Charles. 1968. *A Critical Dictionary of Psychoanalysis*. London: Penguin.

Williams, D. 2006. Why Game Studies Now? Gamers Don't Bowl Alone. *Games and Culture* 1: n.p.

Rust Belt Ruins: The Gothic Genius Loci of Detroit

Leila Taylor

During one particularly brutal Michigan winter in 2015, a pipe burst in the upper floors of a three-story Tudor home in central Detroit. The house was purchased for $70,000 in 2011, but when the owner was unable to keep up with the property taxes, the house was foreclosed, locked, and put up for auction with the family's belongings still inside; furniture, toys, clothes, computers. When the pipe burst, water flooded the house from the top to the basement, cascading out of the windows and freezing into a massive solid waterfall (Stafford 2015). In effect, it turned the house into a giant popsicle earning it the nickname 'The Boston Cooler' after a local treat of Vernor's ginger ale and ice cream. The image is both shocking and beautiful. Each window is obscured and without a view of the interior, this bizarre spectacle is made even more mysterious by our being denied access to the heart of the chaos, but one can imagine pots and pans, armchairs, books, sweaters, and table lamps frozen in space like bits of fruit floating in aspic. Obscure economic structures and environmental forces infiltrate spaces of comfort and stability transforming a home into an object of terror like an eco-horror home invasion (Image 1).

L. Taylor (✉)
Independent Scholar, Brooklyn, NY, USA

H.-G. Millette and R. Heholt (eds.),
The New Urban Gothic, Palgrave Gothic,
https://doi.org/10.1007/978-3-030-43777-0_9

Image 1 Boston Cooler House © Kirthmon F. Dozier/Detroit Free Press /ZUMA Press

We are only as 'safe as houses' if our house is safe and the ruin is, by definition, not fit for human occupation. The Boston Cooler house is no longer a home; it's just a thing that has gotten in the water's way. There is a sublime pleasure in watching this sneak peek of the Post-Anthropocene. Eugene Thacker calls this nebulous zone between now and the future the 'world-without-us' (Thacker 2011, p. 5). It is a glimpse at what our world would be like without people, a place in which human beings are inconsequential. Worse than being ignored, the world doesn't even know we're here. Nature, the true master builder, takes over the design, re-landscaping, reshaping silhouettes, adding filigree where there was none, and eschewing structural integrity reminding us of what we once were and how small we really are. The ruin has no value other than as a spectacle, a metaphor representing our fear of the abandonment of civilization and our powerlessness over nature.

The Gothic has long been the form to express cultural anxieties and as reports on the effects of climate changes become more and more dire, the fictionalized horrors of post-apocalyptic terror become even more personal and more horrifying (Image 2).

(a)

(b)

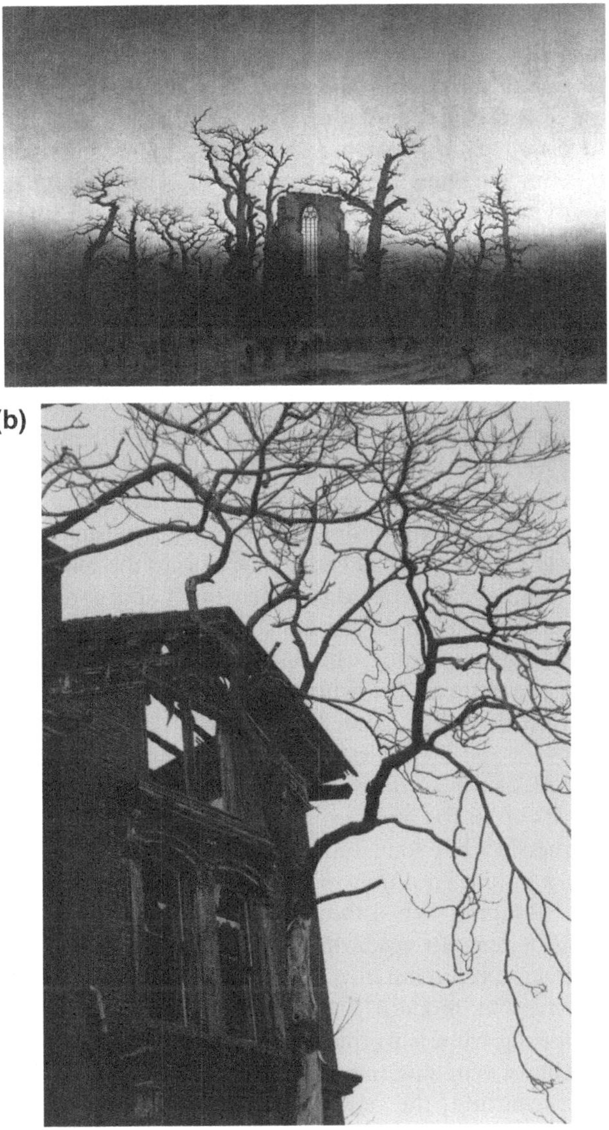

Image 2 Sarah Feinstein, Detroit (c. 1980s)

There is a striking similarity in Caspar David Friedrich's Gothic landscapes and these images of Detroit. Wreckage and decay are part of the visual language of the Gothic landscape: the decrepit castle, the neglected manor, the abandoned Victorian house with creaking doors, crumbling walls, and dripping ceilings, these are classic ingredients of the horror aesthetic. From the ruins of Medieval abbeys scattered across the post-reformation English countryside, emerged a fascination with architectural decay that has ever since been fundamental to the Gothic topography. In the eighteenth century, the architectural folly was an homage to a romanticized Medieval past, and faux war-torn remnants of archways and crumbling castle walls became fashionable lawn decor for the aristocracy. Paintings like *A Ruined Gothic Church beside a River by Moonlight* by Sebastian Pether (1841) and *The Abbey in the Oakwood* by Caspar David Friedrich (1809) romanticized the ruin with misty melancholic nostalgia. In Ann Radcliffe's epic novel *The Mysteries of Udolpho* published in 1794, Emily observes that the castle's 'mouldering walls of dark grey stone, rendered it a gloomy and sublime object … clustering towers, long grass, and wild plants, that had taken root among the mouldering stones' (Radcliffe 2008, p. 226). In Edgar Allen Poe's short story 'The House of Usher', published in 1839, the house's 'principal feature seemed to be that of an excessive antiquity, the discoloration of ages had been great. Minute fungi overspread the whole exterior, hanging in a fine tangled web-work from the eaves' (Poe 2009, p. 34). The device has sustained over the decades and in Guillermo del Toro's film *Crimson Peak* (2015) the walls of Aledale Hall ooze indeterminate goo like open foetid sores and snow piles up in the foyer from a massive hole in the ceiling.

Homeownership has long been part of the fulfilment of the American Dream, a signifier not only of financial accomplishment and stability, but societal belonging. When the owner of the Boston Cooler purchased his house he became an upwardly mobile participant in America's capitalist democracy, a representative of the Manifest Destiny-ism of progress and prosperity. The modern ruin speaks to both the pleasure and the anxiety of bearing witness to the limits of a post-industrial economy and the satisfaction of watching nature's comeuppance existing in a temporal liminality '[permitting] the viewer to see the intact object and its disappearance at the same time' (Steinmetz 2010, p. 316). Like a zombie, the ruin is both alive and dead. In this chapter, I examine three case studies of Detroit as the emblematic Urban Gothic ruin. The stately William Livingstone House, built in 1894 in the wealthy neighbourhood of Brush Park in Detroit, was abandoned and left to decay becoming one of the most photographed ruined houses in the city. Jim Jarmusch uses this same neighbourhood as the location for his 2013 vampire film *Only Lovers Left Alive* revealing the uncanny quality of blight. David Robert Mitchell's 2014 horror film *It Follows* uses the dichotomy between Detroit's suburbia and the city as representations of safe spaces versus monstrous spaces. This chapter explores the Gothic genius loci of Detroit, a city that has become ready-made scenography for horror and a prime example of the New Urban Gothic.

Detroit has become valued for its de-industrialization as much as it used to be valued for its industry and there is a bitter irony in the commodification of the devalued. The fascination with the decay of the constructed environment and the proliferation of photographs fetishizing urban blight, known as 'ruin porn' (Greco 2012), shifts the aesthetics of disrepair and abandonment from the visual language of economic decline and into the realm of romanticism and entertainment. Photographer James Griffioen coined the phrase 'ruin porn' specifically in reference to Detroit in an article for *Vice*, the subversive Canadian magazine turned global media conglomerate, entitled 'Something, Something, Something [...] Detroit' as early as 2009. Tasked with photographing an abandoned school for the magazine, Griffioen observed, 'The photographers are the worst. Basically the only thing they're interested in shooting is ruin porn' (Morton 2009). The images spread and the public appetite for images of the dying city grew. Coffee table books of beautifully photographed abandoned buildings appeared allowing a voyeuristic view of the inaccessible: *The Ruins of Detroit* by Yves Marchand and Romain Meffre (2010), *Detroit Disassembled* by Andrew Moore (2010), and *Lost Detroit: Stories Behind the Motor City's Majestic Ruins* by Dan Austin (2010). These books feature images of a carpet of moss growing on the floor of an empty office, a school gymnasium with wooden floors so warped they look like rolling waves, a knee-high pile of mugshots, wanted posters, and sheets of fingerprints in an old police station. In these frozen moments of ordinary life, the ephemeral becomes permanent and what was thought to be permanent becomes ephemeral.

In 2012 Mark Binelli wrote an article in *The New York Times* titled, 'How Detroit Became the World Capital of Staring at Abandoned Old Buildings' bringing stardom to Detroit's ruins. In considering the meaning behind our fascination with the ruins of antiquity, Binelli writes, 'Ruins don't make you think of the past, they direct you towards the future. The effect is almost prophetic. This is what the future will end up like. This is what the future has always ended up looking like' (Binelli 2012). It is the familiarity and contemporaneity of the modern ruin that creates a dichotomy between attraction and repulsion. We recognize the melted clock on the wall, the broken factory windows, and the piles of school books left to rot. In a temporal flattening of the past, present, and future, we witness our inevitable demise, like looking on your own name on a gravestone.

While the creation of the city is an authoritative, systemized, and logical project, the reduction of the city is chaotic, random, and left to the whim of nature and economics. Detroit was once the fifth largest city in the country and in the early 1900s, the burgeoning auto industry brought a Great Migration of Southerners up North to work in Henry Ford's factory. The 1940s brought a second wave, primarily of African Americans, escaping the segregation of Jim Crow laws. However, blacks were still met with systemic racism and prejudicial housing practices. In the summer of 1967, after decades of police brutality, one of the most destructive and deadly riots in American history broke out (locals call it The Rebellion, not a riot). It lasted five days and left 43 people dead. 1189 people were injured, there were over 7200 arrests, and nearly 700 buildings were destroyed (McGraw 2017).

While the Rebellion was a defining moment in the city's history, there is no single reason for the downturn of Detroit: white flight, corrupt politicians, the economic downturn in the 70s, rising crime rates, but the biggest factor was the decline of the American auto industry. 'The city began to fall apart the minute Henry Ford began to build it. The car made Detroit and the car unmade Detroit. Detroit was built in some ways to be disposable' (LeDuff 2014, p. 80). The city has seen a relentless decline in population, from 1,800,000 in the 1950s to 713,777 in 2010 (WPR 2019). As of 2016, the number was 672,795 (Tanner 2017). Blight is defined as properties that are exposed to the elements, not structurally sound, in need major repairs, have suffered fire damage, or have become dumping grounds and according to a 2014 study by the Detroit Blight Removal Task Force, in 2012 nearly one in three properties in Detroit were considered blighted. This amounted to 380,217 parcels of land, leaving behind a landscape peppered with empty lots and houses swallowed up by vegetation. Pastoral vistas appeared where lines of sturdy middle-class homes once stood side by side.

The night before Halloween, known as Devil's Night, is a holiday unique to Detroit. What started in the 1930s with innocent pranks and mischief (egging houses and toilet papering trees), by the 1970s had grown into a night of vandalism and arson, the worst being October 30, 1984, in which 810 fires were reported throughout the city (Laitner 2015). It took 72 hours to extinguish them all (Chafets 1990). The debauched chaos of Devil's Night has diminished over the years, but Detroit still has the most fires per capita than any other city in America, making it the arson capital of the country. Between 2013 and 2015 alone, more than 10,000 houses, apartments, businesses, churches, and schools burned claiming about 120 lives (Neavling 2016). In the film *The Crow* (1994), the killed and resurrected Eric Draven is seeking revenge for the rape and murder of his girlfriend on Devil's Night. The film ends with the line 'Buildings burn, people die, but real love is forever'.

As a result of all this rioting and burning over the years, buildings that were once flanked by other buildings are left alone on either side turning the neatly planned grid of the city into an irregularly gapped patchwork. As opposed to the prosperity of suburban sprawl, in Detroit, nature is spreading out and taking over where humankind failed (Image 3).

Time and the elements create a specific kind of decay. It's not the decisive human design of vandalism or real estate razing, but the relentless pull of physics and the chaos of the weather. The Livingston House is a beautifully monstrous example of this. Neglect has bent and twisted the building into organic forms with shapes that are unpredictable and unfathomable and it is this titillating 'slippage between what is natural and what is human-made' (Punter and Byron 2004, p. 259) that is both frightening and captivating. Heavier on the left side, the house sags as if it has had a stroke earning it the nickname Slumpy. The rectangular window frames have shifted down into parallelograms. The conical roof of its three-story tower points slightly to the left instead of straight up. The left side, where the front door used to be, leans to the right and the front facade has slipped off revealing a gaping maw. The walls of the first and second floors are gone, exposing the wallpaper of an empty closet inside. It is a structure that, by all rights, should not exist in a modern city and it's a rare treat to witness an object that takes up that much space being allowed to age-naturally. It is no surprise that Slumpy was one of the most photographed ruins in Detroit.

In 1894, William Livingston Jr., publisher of the *Detroit Evening Journal* and former president of the Dime Savings Bank, commissioned architect Albert Kahn to design his residence on Eliot Street in the posh neighbourhood of Brush Park. Designed in the French Renaissance Revival style, the Livingstone House was a classic example of Detroit in the Gilded Age when the city was known as the 'Paris of the Midwest' (Linden 2017). With the advent of the automobile in the early 1900s, people began moving further away from the city centre and single-family mansions were divided into apartments. As the auto industry boomed, freeways sliced through residential neighbourhoods, leading people out of the city. When the auto industry died, Brush Park went from being affluent to derelict. The proud and stately Victorian mansions that used to house society's elites became pathetic and drooping ruins. Around 1987, The Red Cross purchased the land with the intent of demolishing the house to build a new headquarters, but preservationists intervened and succeeded in saving the house by moving the building one block east. Unfortunately, its new foundation was not strong enough to support the

Image 3 The William Livingston House (photo by Kim via Flickr, Creative Commons, 2007)

building and thus began a 20-year slide into decay (Farley and Mullin 2014). Too expensive to either rehabilitate or tear down, the Livingstone House was abandoned to sink under its weight and the houses that used to reside on either side have long since been razed. Slumpy was eventually put out of its misery in the late 1990s and torn down. As dramatic and extreme as this image may seem, it was only one of hundreds of such structures in Detroit.

In Jim Jarmusch's film *Only Lovers Left Alive* (2013) he uses Brush Park, the same neighbourhood in which we find the Livingstone House, as the location for the home of Adam [Tom Hiddleston], a reclusive melancholic vampire. We first meet Adam inside his house perusing a selection of vintage guitars brought to him by his human personal assistant. While shabby and in disrepair, the house is rich and full. There are layers of patterned rugs and drapery in warm jewel tones, a well-worn sofa, walls covered in photographs and paintings of philosophers, poets, and scientists, and musical instruments and recording equipment in every corner. It is a well-lived in space. When we next see Adam he is leaving his house dressed in surgical scrubs on his way to the hospital to meet his contact for a supply of blood. The contrast of the interior and the exterior is jarring. The house is a large decrepit Queen Anne with boarded-up windows painted black. Like the Livingstone House, it is flanked by nothingness. The empty, overgrown lots and the bare spaces on either side are black voids, palpable absences, and there seems to be nothing across the street. It's not the things in the landscape that are ominous; it is the emptiness, the lack of things, the fact that there is a 'landscape' in a residential urban neighbourhood. The location of Adams' house conforms to what Fred Botting attributes to a Gothic landscape. It is in 'an isolated spot, an area beyond reason, law and civilized authority [that empty space] stresses isolation and wilderness, evoking vulnerability, exposure, and insecurity' (Botting 2014, p. 4). The Brush Park of *Only Lover's Left Alive* is a desolate and abandoned place with little streetlight, overgrown weeds, empty lots, and little human activity other than the vampire and an occasional human groupie come to snoop. Instead of the treacherous road to Dracula's castle on a hill, Adam's house is just down the street, yet it is just as desolate and feels just as ominous. The environment is barren and if it is not dead, it's certainly dying. Adam, who is contemplating suicide, is psychologically in sync with the diminishing spirit of the city. But having lived well beyond the scope of a human lifespan, his experience of the world is closer to geological than anthropological. Adam's depression stems from the 'zombies' (his word for human beings) blatant disregard for the earth and the rise of anti-intellectualism. With the apocalyptic urban environment and references to climate change (there are mushrooms sprouting on Adam's lawn months before they are supposed to) the film straddles the line between the Urban Gothic and Eco-Gothic. While the Livingstone House seems at the mercy of the elements, in *Only Lovers Left Alive*, human beings are the monsters wreaking havoc.

When Eve [Tilda Swinton], his partner of eons, comes to visit from Tangier, he takes her on a tour of the city and she notes, 'So this is your wilderness. Detroit'. It is notable that she refers to the city as a 'wilderness'. They drive past notable ruin porn locations: the deserted Packard automotive plant and the Michigan Theater, a majestic French Renaissance showplace now used as a parking garage. The streets are barren and dimly lit as if every other light on the block had been knocked out, a detail that is unfortunately accurate. In 2012, without the budget to repair 88,000 broken streetlights, the city proposed to reduce the number to 46,000 plummeting already crime-ridden and impoverished neighbourhoods in darkness (Davis 2012). In the film, everything is hidden in shadow and it seems to be perpetually three o'clock in the morning. The streets of the Motor City are deserted and the only sounds are the occasional howl of some mysterious creature of the night in concert with a distant police siren.

Unlike *Only Lovers Left Alive*, *It Follows* (2014), directed by David Robert Mitchell, takes place in the suburbs of Detroit. Filmmakers routinely use the American suburb as a location for horror; *Halloween* (1978), *Poltergeist* (1982), *Nightmare on Elm Street* (1984), *Paranormal* Activity (2007). The invasion of the monstrous becomes even more terrifying against the background of perceived financial security and cultural hegemony. The protagonist in *It Follows*, Jamie 'Jay' [Maika Monroe] is a high school student living in pleasant middle-class, neighbourhood in the suburbs of Detroit. We see children playing in front of nice houses, a man mowing his lawn, and a handsome teenager washing his car. It's not a wealthy neighbourhood, but is almost idyllic in its peaceful normalcy. When we first see Jay she is floating lazily in an above ground pool in her back yard before going on a date to the movies with her new boyfriend Hugh [Jake Weary]. The date ends quickly when Hugh becomes frightened by a vision of a woman, something Jay cannot see, and insists they flee the theatre.

The next evening, we see Hugh and Jay holding hands walking through a forest with just a hint of a building between the branches. The scene suddenly shifts to the couple on a beach and while we know it is the same evening, the change in location is disorienting. The scene shifts again to a deserted parking lot in front of a massive abandoned institutional building where they have sex in their car. Afterward, Hugh chloroforms Jay into unconsciousness, and when we next see her, Hugh has tied her to a wheelchair, presumably inside the building on the second or third floor. It's a vast and empty space of crumbling concrete, exposed rebar, and missing exterior walls entirely exposing the view of a dense and wild forest. Hugh explains to her that after having sex with him she is now the carrier of a sexually transmitted curse in the form of a human-looking shape shifting entity visible only to the targeted person. Hugh tells her that until she passes the curse on by having sex with someone else, the creature will keep following her until it kills her.

For this pivotal scene Mitchell uses two different abandoned buildings as locations: the exterior is the Northville Psychiatric Hospital located just outside of Detroit and the interior is shot inside the abandoned Packard Automotive Plant on the East side of the city. Opened in the 1952, Northville Hospital fell victim to the recession in the 70s resulting in significant overcrowding and neglected patients. The hospital eventually closed in 2003 and has since become a popular destination for intrepid ghost hunters. The woods surrounding the hospital, presumably where Hugh and Jay were walking, are known as the Evil Woods. The 43-acre Packard automobile manufacturing plant, opened in 1903, and was considered the most modern factory in the world at the time. After its closure in 1952, it was left to ruin and became the largest abandoned building in the world (Kwong 2016). It's easy to pass the film off as another cautionary tale against STDs, but Mitchell's use of these ruined institutions and the contrast between the city and the suburbs suggests a different kind of anxiety, one centred on economics and race instead of sex.

Jay and her friends seek out Hugh for answers and learn that he has fled the safety of the suburbs and is hiding out in a decrepit abandoned house on a dismal street in Detroit. As they drive into the city, they pass row after row of boarded-up shops, dilapidated houses, and empty lots. When Adam and Eve (the vampire couple from *Only Lovers Left Alive*) take a drive through the city, it is portrayed with a kind of philosophical contemplation and a historical perspective. By contrast, in *It Follows* the trip is depicted as a dangerous journey, marked with dread as the film continues to shift from the suburbs to the city during key points of action. When they try to run and hide from the entity, they flee to a friend's summer cottage by the beach. When they decide to confront the entity and attempt to destroy it, the location they choose is a public pool away from the suburbs in Detroit. As Jamie and her friends march back into the city for this battle, it is with defiant gravitas and the atmosphere of an epic quest. There are few people on the streets and unlike their neighbourhood in the suburbs, they are all African American. As they walk past boarded-up houses and lawns overrun with weeds one character says:

> When I was a little girl my parents never let me go south of 8 Mile [the road considered to be the border between the city and the suburbs] and I never knew what that meant until I got a little older and I realized that's where the city started and the suburbs ended. (*It Follows*)

Euphemisms like 'the inner city' and 'urban' have long been codes for blackness. Discriminatory practices in the 1930s reduced housing options for black people and the subsequent ghettoization to poorer, less maintained and derelict neighbourhoods made urban areas undesirable. As the automobile flourished, whites moved further out into the suburbs taking their money and property values with them. As drugs infiltrated the city and crime rose, black neighbourhoods (and people) became associated with violence and poverty. For Jamie and her friends, to venture into Detroit meant going into a majority black environment, one that white America has been taught to fear unequivocally. The fear of the entity, the 'it', becomes fear of crime, poverty, and black people (Kelly 2017). As an 'infected' person, Hugh isolates himself in an urban quarantine away from the security of suburbs. When the group attempts to destroy the creature, the plan is carried out in the city establishing Detroit as a monstrous place accustomed to neglect and violence; a place to kill and a place to die.

An unusual detail of *It Follows* is that time is ambiguous. The decor of Jamie's house seems to be from the 70s, yet her friend reads a book from some electronic device not yet invented. For her date with Hugh, the couple go to a silent movie and their friends watch cartoons from the fifties on televisions from the 80s. Jamie swims in an outdoor pool, yet it seems to be autumn. It creates a disorienting effect, being simultaneously in the past and the future, but this nebulous time frame, is one that seems suited to the city. Detroit is continually living in an ebb and flow of recovering from the past and imagining the future: the rise of the auto industry and its fall, the mass migration of blacks from the South followed by the mass exodus of whites to the suburbs.

In the documentary *Detropia* (2012), film makers Heidi Ewing and Rachel Grady show us the real people who live and work in Detroit, not just the empty buildings that surround them. Crystal Starr, a video blogger and community activist documents the political strife of her city and takes the documentarians into ruined spaces, not to gawk, but to reminisce about her childhood and lament over the wasted opportunities in these abandoned buildings. We see her at work behind the counter in a diner when a pair of tourists from Switzerland comes in for coffee. She asks what brought them to the city and they reply, 'Everything in Switzerland is so neat and proper and new and Detroit seems kind of interesting with all its decay'. I'm bothered by sentiments like this, but I admit I'm also fascinated by the ruinscape. There is a guilty pleasure in these images (hence the 'porn' connotation), but having grown up in Detroit, it's uncomfortable watching my city's weaknesses exploited. There is a dissonance between my fascination with these images and the circumstances in which they are made. The ability to delight in the ruin is a privileged position and the spectacle of annihilation is only pleasurable when you're not the one being annihilated. The aesthetics of decay can easily fall into the trap of romanticizing poverty, but these images are the result of economic decline, political corruption, systemic racism, and violence. It dismisses the people and the communities who have been there all their lives and ignores positive signs of growth. There has been a reverse migration of young entrepreneurs moving to Detroit from around the country, including Roslyn Karamoko founder of Détroit Is the New Black, a boutique clothing store with the goal to help 'Detroit move forward while ensuring that those who have been there all along still have a place at the table and the means to stay there' (Battis 2017). In *Only Lovers Left Alive*, Eve sees the ruins as a moment in the city's history and prophetically expresses the city's resilience, 'This place will rise again. There's water here. When the cities in the South are burning, this place will bloom'.

Detroit has a few nicknames (The Paris of the Midwest, Motown, The Motor City), but it is also known as The Renaissance City, dying in flames only to be reborn anew, over and over again. The abandoned industrial buildings are rehabilitated as artist studios, the patches of land left by arson are transformed into community gardens, and in 2018 The

Ford Motor company purchased the massive, long abandoned, Michigan Central Station, the quintessential icon of Detroit ruin porn. In January 2017, 88,000 dead streetlights were replaced with 65,000 brand new LED lights (Reindl 2016). Brush Park, where The Livingstone House (aka Slumpy) once stood and where Adam made his home, is currently undergoing a massive redevelopment, filling in the awkward gaps of abandoned lots with sustainable LEED certified houses, restaurants, and shops. Most recently Detroit has been voted 'one of the hottest cities to visit' by the travel guide *Lonely Planet* and in a representation of commodified Urban Gothic, the renovated home of Adam the vampire was, for a while, available to rent as an Airbnb.

References

Austin, D. 2010. Michigan Theater. historicdetroit.org.

Battis, L. 2017. Here's the Motor City's Hottest Designer: "Détroit Is the New Black". *Time: American Voices*. time.com.

Binelli, M. 2012. How Detroit Became the World Capital of Staring at Abandoned Old Buildings. *New York Times Magazine*, 12 November.

Botting, F. 2014. *Gothic: The New Critical Idiom*. London: Routledge.

Chafets, Z. 1990. The Tragedy of Detroit. *New York Times Magazine*, 29 July.

Davis, L. 2012. Detroit Plans to Shrink by Leaving Half the City in the Dark. 26 May. io9.gizmodo.com.

del Toro, G. (dir.). 2015. *Crimson Peak*. Legendary Pictures.

Detroit Blight Removal Task Force. timetoendblight.com.

Ewing, H., and G. Rachel (dir.). 2012. *Detropia* (Loki Films).

Farley, R., and J. Mullin. 2014. Brush Park Historic District/Woodward East Historic District. Detroit1701.org.

Greco, J. 2012. The Psychology of Ruin Porn. 6 January. citylab.com.

Griffioen, J. 2009. Schools Out Forever. 1 February. vice.com.

Jarmusch, J. (dir.). 2013. *Only Lovers Left Alive*. Sony Pictures Classic.

Kelly, C.R. 2017. *It Follows*: Precarity, Thanatopolitics, and the Ambient Horror Film. *Critical Studies in Media Communication*. https://doi.org/10.1080/15295036.2016.1268699.

Kwong, M. 2016. Meet Al Hill, the Sole Resident of the World's Largest Abandoned Building. *CBC News*, 28 February. cbc.ca.

Laitner, B. 2015. Halloween Week Fires Down in Detroit This Year. *Detroit Free Press*, 1 November.

LeDuff, C. 2014. *Detroit: An American Autopsy*. New York: Penguin Books.

Linden, C. 2017. CuriosiD: Why Is Detroit Sometimes Called "The Paris of the Midwest?" 13 September. WDET.org.

Marchand, Y., and M. Romain. 2010. *The Ruins of Detroit*. Göttingen: Steidl.
McGraw, B. 2017. Detroit '67: By the Numbers. *Detroit Free Press*, 22 July.
Mitchell, D.R. (dir.). 2015. *It Follows*. Northern Lights Films.
Morton, T. 2009. Something, Something, Something… Detroit. 31 July. Vice. com.
Neavling, S. 2016. How 10,000 + Fires Devoured Detroit Neighborhoods over the Past 3 Years. 17 March. motorcitymuckraker.com.
Northville Regional Psychiatric Hospital. michiganhistory.leadr.msu.edu.
Poe, E.A. 2009. *Great Tales and Poems*. New York: Vintage Books.
Proyas, A. (dir.). 1994. *The Crow*. Miramax.
Punter, D. and G. Byron. 2004. *The Gothic*. Malden, MA: Blackwell.
Radcliffe, A. 2008. *The Mysteries of Udolpho*. Oxford: Oxford University Press.
Reindl, J. 2016. Detroit Streetlights Go from Tragedy to Bragging Point. *Detroit Free Press*.
Stafford. K. 2015. Boston Edison Home Frozen Solid After Pipe Bursts. *Detroit Free Press*, 12 January. freep.com.
Steinmetz, George. 2010. Colonial Melancholy and Fordist Nostalgia: The Ruinscapes of Namibia and Detroit. In *The Ruins of Modernity*, ed. J. Hell and A. Schönle. Durham and London: Duke University Press.
Tanner, K. 2017. Detroit's Population Still Down, Despite Hopes. *Detroit Free Press*, 25 May.
Thacker, E. 2011. *In the Dust of This Planet*. Washington: Zero Books.
World Population Review. 2019. Detroit Population. 30 March. worldpopulationreview.com.

Global Gothic: Decentring the Urban Gothic

Ruth Heholt

At the 2018 British Association of Victorian Studies (BAVS) conference in Exeter UK, the papers went global. The term 'Victorian' has, for a long time and in some of the central discussions of the era, often come to stand in for the time period spanning the nineteenth century. 'Victorian' must of course, refer to the reign of Queen Victoria in Britain, but it has morphed and expanded to sometimes mean the whole of the 1800s and is a term applied to several countries' own experience of the nineteenth century: cultural production, social constructs and representations in art and literature. The main discussions at the conference examined the idea that 'Victorian', perhaps defying expectations, did not just reside with Britain, but formed a multiple, but at least somewhat cohesive movement centring around the term 'Victorian' that extended across the globe. The colonial and post-colonial implications of such a thesis were of course clear. That India, America, Canada, parts of Africa, the Caribbean and other colonised places might have a 'Victorian era', given the extent of British colonial rule, is perhaps to be expected. Although, understandably, each of these 'Victorians' is going to be different, multiple, unexpected and locally inflected. However, the twenty-first-century global scholarly community still use the term 'Victorian'. The flow of ideas, discourses, representations, morals, social mores and customs apparent in the nineteenth century, while showing marked and important differences in different locations, were still, in the discussions at the conference, identified as 'Victorian'. This was not an imposition from the delegates, but more of a self-identification that originated from many different global locations. Thus, and for example, there is NAVSA (North American

Victorian Studies Association) and AVSA (Australasian Victorian Studies Association). However, rather than seeing a centrally located 'real' Victorian, many of the papers and keynote addresses looked at global networks and intersections; filigree tentacles of texts and ideas that flowed in various directions and which interweaved with each other, sometimes regardless of the 'centre'. There were, in effect, different Victorian*s* emerging in different spaces and places. And what might be described as this 'transnational turn' in Victorian studies can (and is being) replicated in Gothic studies.

In 2002, the blurb on Avril Horner's book *European Gothic* cites 'the tyranny of the Anglo-American narratives that have dominated critical histories of the Gothic'. The usual story of the emergence of the Gothic novel is that it begins in Britain in the eighteenth century. Angela Wright (and very many others) cite the Gothic's 'beginnings in 1764 with Walpole's *The Castle of Otranto*' (2007, p. 1). Historically the 'origin' texts have been discussed as being British: with Ann Radcliffe, William Golding and Matthew Lewis heading the list. This is what Andrew Smith has termed a 'Gothic style' (2007, p. 2). Smith argues that this style 'was given new impetus in the mid-eighteenth century with the emergence of Enlightenment beliefs that extolled the virtues of rationality. Such ideas', Smith says, 'were challenged in Britain by the Romantics' and he argues that the Gothic 'often shares in such anti-Enlightenment ideas' (2007, pp. 2–3). Smith problematises this, (saying for example that early Gothic texts were 'often highly formulaic' (2007, p. 3), but still, the texts first cited are British ones. In fact, the Gothic tradition in fiction is usually expanded out to share its origins with American Gothic. Thus, the recent (and excellent) book *Neoliberal Gothic* (2017) cites, in the introduction, *The Castle of Otranto*, Charles Dickens, Robert Louis Stevenson, Bram Stoker and Oscar Wilde before moving on to writers from the United States: Washington Irving, Nathanial Hawthorne, Stephen King and Edgar Allan Poe (Blake and Soltysik Monnet 2017, pp. 2–3). The collection does, however, explore 'literary, televisual, filmic and dramatic works from countries as geographically distant and culturally diverse as Britain and Mexico, the United States and Ireland, Germany, Korea, and Japan' and states that these Gothic works 'have both a nationally specific context, and a global awareness' (2017, p. 3). Indeed, this does signal the transnational turn of the global Gothic, but the discussion still begins with British and American texts. Linnie Blake and Soltysik Monnet claim that as writers in the United States 'had drawn on the symbolic machinery of the eighteenth-century English Gothic to illuminate their own concerns ... [t]his pattern of gothic writing interrogating the contemporary *Zeitgeist* would spread across the world ... as successive nations adopted the mode's lexicon to explore their own troubling times' (2017, p. 3). This explains what might be called the 'rise of the Gothic' as concentric circles, originating and rippling out from Britain (or here, more narrowly, England).

Horner says that her book *European Gothic* 'offers a corrective to what has become a somewhat lopsided story of the development of the Gothic novel. All too often academic accounts of the Gothic repeat, almost mantra-like, the names at the core of the Anglo-American tradition: Walpole, Radcliffe, Lewis, Shelly, Brockden Brown, Hawthorne, Poe' (2002, p. 1). 'Where', she asks, 'are the comprehensive histories of French or Italian Gothic for example?' (p. 1). Horner's critique is understandable and there are of course, immediate caveats to this 'mantra' of canonical Gothic texts. However, I am not accusing Smith, Wright, Blake or Soltysik Monnet of being Anglo-centric. The point they (and others) make is that it is these texts and the work of specific authors that were the first *characterised* as Gothic as it was turned into an aesthetic movement and style. This idea of characterisation (or one might say, labelling) is important as the 'Gothic' itself evolves from elsewhere and another context entirely. Of course, the Goths themselves did not come from Britain: they were a Germanic tribe that existed from the third to the fifth century C.E. At the time of writing, the Gothic architectural gem of Notre Dame is still smouldering and has barely been saved from a near-devastating fire, bringing this type of architecture to the fore of international consciousness. Although this is not the place to discuss Gothic architecture (mostly because it is outside of my field of expertise) it is worth noting that in the first place, this style of building has been around for hundreds of years, and secondly that it is European in origin, (again not just British). Indeed, relative to Gothic novels, a vital part of this 'style' involved the settings of the first British Gothic novels, and very many had European locations: Italy, Spain and Germany for example. Thus, even from what one might term an Anglo-centric point of view, the Gothic is at the very least, European and American rather than more narrowly British.

If the Gothic is more-than-British, then Britain itself was expansionist and, with Imperial aggression, had invaded and colonised many parts of the world. Smith and Hughes in *Empire and the Gothic* state that there is a 'link between the Gothic and the colonial' (2002, p. 4). Colonial invasion (physical, metaphysical and psychic) was conducted, as Patrick Brantlinger has evocatively suggested, under a 'Rule of Darkness' (1988). This darkness itself gave rise to the Gothic. Alexandra Warwick identifies what she calls 'Colonial Gothic' which in its away-from-Britain mode sees a strand of literature 'dealing with the experience of the coloniser abroad' (1998, p. 261). Here she says 'both landscape and people are seen as uncanny, beyond the possibilities of explanation in European terms' (p. 262). However, this is still in Euro-centric mode, where Orientalism, and European fears of the Other hold sway. As Smith and Hughes note, these types of (usually travel) 'narratives [capture] the European fear of its contact with the Eastern Other' (p. 6). This 'other' is identified as foreign—not-British—and is frightening, threatening and atavistic. Marie Mulvey Roberts contends in *Dangerous Bodies* that the Gothic 'depends upon the consensual formation of a monstrous alterity, whether it be vampire, ghost, demonic stigmatic or man-made monster' (2016, p. 3). She continues, 'it is the very idea of the monster that sustains social, economic and sexual hierarchies. The Gothic monster has been the rallying point for cultural, nationalist or religious hegemonies' (pp. 3–4). This is the making of monsters from a position of power, albeit unstable and fractured, that turns those bodies deemed 'foreign' and 'other' into freaks and beasts. But these terrible creatures are less a reality of the other than a mirror reflection of the fears and anxieties of those who reside in the Western centres. Smith and Hughes note that, 'One of the defining ambivalences of the Gothic is that its labelling of otherness is often employed in the service of supporting, rather than questioning, the status quo' (2003, pp. 2–3). Making those who are seen as 'other' monstrous; playing out colonial fears through the 'foreign' body or location can, of course be seen in this way. That which is not 'us' must be de-humanised and eradicated. However, while this colonial construction of the Other from its own specific point of view can lead to both a re-creation (and perhaps affirmation) of Imperial values and viewpoints, at the same time it can radically destabilise the underlying suppositions, assumptions and impositions of the colonial point of view. The monster, at the very least, looks back providing a distorted mirror through which the coloniser sees himself and, in fact, nothing else.

This equates with the conception of the post-human as well as the Gothic. Anya Heise-von-Lippe, in *Posthuman Gothic* notes that the notion of the post-human 'challenges the basic paradigms we associate with being human' and shakes the centrism 'which ties Western thought to an anthropocentric perspective often perceived as universal' (2017, p. 2). This idea of viewpoint is crucial in the discussion of Global Gothic. Whose point of view is or can be seen? From which part of the globe is the viewpoint taken? If taking the view of the colonised, there are historical examples of violently created Gothic modes arising from Imperial and colonial violence and aggression. Thus, for example, from Haiti the figure of the zombie which, Kyle Bishop argues, is 'a creature born of slavery, oppression, and capitalist hegemony' (2010, p. 37). And Glennis Byron cites 'soucouyants, La Llorona, pontianaks and *onryō*' (2013, p. 7) as being the focus of specific Gothics from around the world. Thus, the Gothic has roots in different parts of the world which must mean that there are Gothics that are away from Britain. This type of colonial Gothic though is enforced onto the local populations' folklore and fiction, birthed through violence and oppression. However, locations across the globe were developing their own Gothic modes, even if these only came to be labelled 'Gothic' later. In the important text *Globalgothic*, Byron looks at new manifestations of the Gothic around the world. While focussing on 'the late twentieth and early twenty-first centuries' (2013, p. 1), she nevertheless makes forays into new territory, looking at 'multidirectional exchanges' (2013, p. 3) that do not have one central place of origin. Byron states that 'the tropes and strategies Western critics have associated with the gothic, such as the ghost, the vampire and the zombie, have their counterparts in other cultures, however differently these may be inflected by specific histories and belief systems' and says that she wants to 'register a sense of a gothic inextricable from the broader global context in which it circulates rather than a gothic tied to past notions of Enlightenment modernity' (2013, pp. 3–4). Byron opens things out (that are sometimes closed again consequently), and points to 'a need to decentre the West when considering globalgothic' (2013, p. 4). Some of Byron's collection examines globalisation itself as encompassing gothic modes and motifs, but other essays look at Gothic texts that originate away from the centre. Byron coins a term I find to be particularly useful: 'Indigenous gothic' (2013, p. 6) and it is this sense that there are Gothics that come from and have origins *elsewhere* (from a Western point of view) that is most important here. Different places produce different Gothics these do not necessarily have anything to do with either Britain, Europe or the United States.

For the purposes of this volume and the essays in this section, it is the Gothics originating from global cities that are of interest. As Robert Mighall says, 'The relocation of the Gothic in the modern city involved the city itself, or at least part of it, being Gothicized' (1999, p. 31). Here the darkness, decay, hidden secrets and perils are manifested in different and specific ways in different cities around the globe. Each city in the chapters that follow creates a different sense of the Gothic to that of those located elsewhere. From Singapore to New Delhi, from Srinigar to London and on to Barcelona, different forms of the Gothic are birthed, originate and reside in these places. Yet in true global and Gothic fashion, boundaries are breached and commonalities and cross-fertilisations of the Gothic mode, style and motifs are evident and growing. One of the aims of this whole collection is to destabilise and decentre the 'canonical' Gothic cities, most notably perhaps London. The essays that follow examine how marginalised urban communities are represented and how Gothicised violence and injustices like poverty, colonisation and war haunt the urban landscapes. Yet, despite these traumas and atrocities, (or perhaps even because of them), at the same time, the idea of a Global Gothic can provide some space for empathy and connection.

References

Bishop, Kyle. 2010. *American Zombie Gothic: The Rise and Fall (and Rise) of the Walking Dead in Popular Culture*. Jefferson, NC: McFarland.

Blake, Linnie, and Agnieszka Soltysik Monnet (eds.). 2017. *Neoliberal Gothic: International Gothic in the Neoliberal Age*. Manchester: Manchester University Press.

Brantlinger, Patrick. 1988. *Rule of Darkness: British Literature and Imperialism, 1830–1914*. Ithaca, NY: Cornell University Press.

Byron, Glennis (ed.). 2013. *Globalgothic*. Manchester: Manchester University Press.

Heise-von-Lippe, Anya (ed.). 2017. *Posthuman Gothic*. Cardiff: University of Wales Press.

Horner, Avril (ed.). 2002. *European Gothic: A Spirited Exchange 2760–1960*. Manchester: Manchester University Press.

Mighall, Robert. 1999. *A Geography of Victorian Gothic Fiction: Mapping History's Nightmares*. Oxford: Oxford University Press.

Mulvey-Roberts, Marie. 2016. *Dangerous Bodies: Historicising the Gothic Corporeal*. Manchester: Manchester University Press.

Smith, Andrew. 2007. *Gothic Literature*. Edinburgh: Edinburgh University Press.

Smith, Andrew and William Hughes (eds.). 2003. *Empire and the Gothic.* Basingstoke: Palgrave Macmillan.

Warwick, Alexandra. 1998. Colonial Gothic. In *The Handbook to Gothic Literature*, ed. Marie Mulvey-Roberts, 261–262. New York: New York University Press.

Communal After-Living: Asian Ghosts and the City

Katarzyna Ancuta

Cities are environments designed by and for humans. In the age of the Anthropocene, when human activity has been recognised as a force capable of reshaping geological landscapes and determining the livability of the planet, urbanisation rapidly becomes equivalent with the human condition, and the urban mind—habits of thinking developed as a result of city life—becomes 'a concept of normalcy' (Fisher and Herschend 2010, p. 197). The urban mind aims to alleviate potential threats to urbanity by remodelling urban spaces, social norms and interactions, and redefining its attitude to the non-urban—a broad category inclusive of such qualifiers as the rural, the non-human or the supernatural (p. 198). Framed by its socio-historical context and reflecting changing ideologies, this attitude is always culturally specific. In Asian cultures, shaped to a large extent by underlying animistic beliefs and practices that continue to inform popular interpretations of official religious and philosophical doctrines in the region, ghosts and humans have always been part of the same universe. If the urban condition is to represent normalcy, then Asian cities need to accommodate their ghosts alongside the living.

K. Ancuta (✉)
Faculty of Arts, Chulalongkorn University, Bangkok, Thailand

H.-G. Millette and R. Heholt (eds.),
The New Urban Gothic, Palgrave Gothic,
https://doi.org/10.1007/978-3-030-43777-0_10

Ghosts in Gothic have long been theorised as the residue of the past, the return of the repressed or the aftershock of personal and collective trauma. Andrea Juranovszky writes of the 'Gothic loop' where the return of ghosts demonstrates that '[t]he Gothic present can only be imagined in terms of an endless repetition of the same traumatic moment' (2014, p. 3). Avery Gordon reminds us that the ghost is 'a social figure, and investigating it can lead to that dense site where history and subjectivity make social life' (2008, p. 8). Haunting is also always inevitably linked to a place, but then in the world shaped by humans '[t]here is no place that is not haunted [...] Haunted places are the only ones people can live in' (de Certeau 1988, p. 108). Asia is by far the most populated continent—there are currently more people living in Asia than on all the remaining continents combined—and the most densely inhabited areas of Asia are its cities. Driven away from the countryside following rampant deforestation, excessive river regulation and accelerated urbanisation, Asian ghosts have adapted to living in these cities, embraced new technologies and infiltrated new spaces. This chapter discusses one such space—the multi-apartment building designed to house a large number of tenants.

A haunted apartment/apartment complex stands in an opposition to a haunted house. The haunted house is 'instantly recognizable [...] identified as a troubled place, marked by neglect, strange habits and failed rituals of order and maintenance' (Curtis 2008, p. 31). Excluded from the bustling human community it clearly does not belong to, the haunted house presents its explorers with a set of thresholds that should not be crossed. Its foreboding exterior hides an impossible interior that seems to expand or contract on a whim. There are always too many doors and staircases leading nowhere in particular; cellars, attics, utility rooms and closets transform into hidden passages leading to secret locations; strategically placed mirrors create a maze of reflections and offer glimpses of other dimensions; while gloomy family portraits of previous owners stare ominously at anyone brave enough to venture inside. Unlike the haunted house, the haunted condominiums, high-rises and tower blocks, apartment houses and housing estates represent the mundane and the ordinary. If houses are haunted by the memories of their previous owners, apartments are often rented out on short contracts and their tenants have less hold over the place. Houses offer a clear division between the inside and the outside, for apartments this distinction appears less pronounced. Living in an apartment building tenants are simultaneously invisible and isolated, and belong to a greater community. Although in horror, apartments can also appear structurally unstable and prone to transformation, their interior design privileges simplicity and minimalism. The exteriors of the buildings similarly do not draw attention to themselves.

The chapter looks at three distinct themes in what we could loosely term Asian 'apartment horror' films that are characteristic of the specific sociocultural contexts and urban cultural economies they represent: (1) the portrayal of the contiguous community where ghosts co-habit the space alongside the living; (2) the alienating character of modern urban communal lifestyles, where ghosts are more visible than the living, and the biggest fear of both groups is that of loneliness and isolation; and finally (3) the placement of the ghost as a representation of a failed dream of economic success that continues to drive the migration of Asian rural populations to the cities. The first theme is characteristic of the ghost films made in Hong Kong, Singapore and Taiwan—culturally Chinese territories that have escaped the Chinese Communist Party's suppression of supernaturalism. Imposed during the Cultural Revolution, the ban on religion, seen as incompatible with the materialist ideology, eliminated traditional forms of spiritual practice and exorcised ghosts from contemporary cultural production in mainland China. With the 'wronged spirits and violent ghosts, monsters, demons, and [...] strange and supernatural storytelling' deemed ideologically offensive, the Chinese government has remained committed to cleansing 'the negative effect these items have on society [...] and to protect adolescents' psychological health' until today (Reuters 2008). Unsuprisingly then, Chinese ghost films are mostly made in Hong Kong.

Themes of loneliness and urban isolation, discussed in the second section, permeate ghost films made in Japan and South Korea, two highly developed market-oriented Asian economies, whose cities are exemplaries of neoliberal urbanism developed in accordance with the dynamics of global financial capitalism. The implementation of neoliberal urban planning policies has contributed to the reorganisation of such cities generating new spatial and social relations and dissolving traditional community ties. The final part of the discussion focuses on Bangkok, as an example of a primate city—a disproportionately large urban agglomeration that dominates the entire country, which represents a model characteristic of most Southeast Asian economies. The appearance of ghosts in such cities can be seen as part of an ongoing migration of the countryside population into the central urban area, although ironically, migrant ghosts are more visible than the living.

Contiguous Communities: Living with Ghosts

Asian ghost narratives are concerned with portraying complex relationships between the living and the dead. These relationships are based on two general assumptions—that ghosts retain their anthropomorphic form, which potentially makes them indistinguishable from humans; and that they are always present among us. Common plots feature ghosts that fall in love with humans, enter into sexual relationships, give birth to children, take care of their loved ones, avenge their misfortunes, reveal secrets, punish wrongdoers, attempt to reincarnate or resurrect themselves, or, occasionally, try to annihilate all humankind. For the majority of these plots to succeed, humans need to remain largely unaware that they are interacting with ghosts. As the earliest such stories originated in China,[1] it seems fit to begin the discussion with Chinese ghosts, and the most engaging cinematic representations of such ghosts in Hong Kong.

To say that Chinese afterlife is complicated is an understatement. Chinese ghosts appear trapped between three flexibly interlocking belief systems 'caught up in Buddhist karmic cycles, pursuing Confucian ideals and bound by Taoist rituals' (Lim 2005, p. 8). Chinese religions have a notoriously pragmatic view of both this world and the next, concerned mostly with various aspects of prosperity, such as health, wealth and descendants (Berling 1992, p. 184). The afterlife is expected to be a mirror image of the earthly domain (with Heaven and Hell structured to resemble the Imperial Court and its prison), and ancestors are sent off to the next world accompanied by bundles of offerings meant to safeguard their comfortable existence after death (p. 188). Standard offerings to ghosts include various types of food, paper money, clothing and personal items, while ancestors additionally receive elaborate paper houses with effigies of servants to maintain them, cars, electronics and other luxury items, and infernal currencies of higher denomination to bribe the officials of Hell, if needs be (Image 1).

The Confucian value system ensures that the living are obligated to care for their ancestors. Ghosts of those who died without descendants, whose sudden deaths prevented them from returning home, or who did not receive a proper burial are more likely to roam the Earth unsatisfied.

Image 1 A selection of paper offerings for the dead

Barbara Ward and Joan Law conclude: 'All these are the underprivileged dead. They get none of the food, paper clothing, and *spirit money* that are showered upon *ancestors,* and none of their great respect. And they resent it' (cited in Scott 2007, p. 92, emphasis original). To put such ghosts at ease, the living need to acknowledge their existence and ensure they have no reason to haunt them: 'ghosts who are given offerings do not have to be feared. For they are satisfied and do not harm anyone' (Scott, p. 95). The lingering ghosts depend on the living for assistance. Unsurprisingly then, they tend to flock towards human communities, and as family ties become less pronounced and opportunities to die away from home increase, earthbound ghosts are drawn to large urban agglomerations, just like the rural poor migrating in search of a better life.

While not all the ghost films made in Hong Kong are set in the city, those that are tend to evoke the aesthetics that reveals a dark side to Hong Kong's economic success and financial prosperity. The hungry ghosts roaming the city rarely make it to its glamorous side. Their Hong Kong is the city of 'sweatshops, storefronts, urban pollution, and shantytowns of unrelieved squalor, with too many people and too little land' (Hoover and Stokes 1998, p. 25). Just like the living, the ghosts are the by-product of the city's 'cutthroat' capitalism. In Chinese beliefs, ghosts roam the Earth if they have been denied a proper burial, however, the scarcity of land in Hong Kong has made an ownership of a grave yet another luxury few can afford. Traditionally, in Chinese culture the body is meant to be buried in the ground near one's place of origin. Since 2013, more than 90% of Hong Kong's dead have been cremated instead. Even then, an average waiting time for a place in a columbarium is currently four years and the authorities estimate that by 2023 the city will be facing a shortage of 400,000 urn spaces. Those who insist on being buried are expected to be exhumed after just six years. Other options include a sea burial or having one's ashes scattered from a plane (Blundy and Davis 2017). Things are not much better for the living, as the limitation of space engenders a cancerous form of architecture with the buildings sprouting ever smaller units within to accommodate a growing number of people living under the poverty line. Apartments are often further divided to include a number of tiny cubicles, aptly named 'coffin homes', where the poorest of the poor live out their ghostly lives. Sadly, their fate has little chance of improving in the afterlife.

Housing estates are common haunted spaces in Hong Kong films. *Rigor Mortis* (2013), the directorial debut of the singer/actor Juno Mak, set almost entirely in a run-down high-rise apartment complex, offers an interesting depiction of an urban ecosystem where humans and ghosts coexist in relative harmony. The Cantonese title of the film, *Geung si*, refers to its main antagonist—a hopping vampire, or *jiangshi* (in Mandarin)[2]—but the film features a whole array of local ghosts. The film tells a story of a retired actor, Siu-ho Chin, who rents out an apartment to commit suicide. He is rescued by a local resident, Yau, a former Taoist exorcist who runs a small restaurant in the building. Yau feeds both the living and the dead. When reminded that 'Auntie Lai passed a month ago', he retorts: 'So? She still needs to eat'. After a black magic practitioner, Gau, resurrects one of the tenants as a vicious vampire, Siu-ho joins Yau in his fight against evil. The ensuing battle claims the lives of all the protagonists but the *jiangshi* is eventually defeated.

The dilapidated condition of the building matches that of its inhabitants—the elderly, the forgotten and those who wish to isolate themselves from the world. Indeed, it seems as if the world does not exist beyond the concrete walls of the high-rise inhabited by both the living and the dead. Siu-ho's apartment contains two angry *nu gui*—vengeful spirits of twin sisters who killed themselves following a sexual assault. Gau shares his living quarter with scores of *xiao gui* baby spirits he raises and consumes to alleviate his cancer. Solemn processions of tall ghosts under umbrellas patrol the hallways in search of souls they are meant to escort to Hell. Known as the soul collectors they are said to use their umbrellas to contain the defiant ghosts. Their recurring appearance is likely connected to Gau's terminal disease but as he continues to cheat death they leave empty-handed. Mirrored surfaces reflect the ghosts of the deceased tenants who linger around the building with nowhere to go. Last but not least, there is also the titular *jiangshi*—a reanimated corpse of the building's caretaker killing people and creating more ghosts.

The closing sequence of the film takes us back to the beginning. We are once again shown Siu-ho arriving at his apartment and all the tenants seem very much alive. This time, however, Yau is late and Siu-ho succeeds in hanging himself. The circular narrative suggests that perhaps all the characters in the film are already ghosts, a plausible explanation since each protagonist seems to correspond to a known local ghost type. We can identify the main protagonist as *diao si gui*–or a ghost of a person who died by hanging; the mother who refuses to leave the building is *di fu ling*—a ghost tied to the place of her death; her always hungry albino son is likely an *e gui*—a hungry ghost with pale grey skin doomed to suffer insatiable hunger; the clothes-mending friendly old lady who resurrects her husband as a monster is a *gui po*—an old woman ghost that has good intentions but can occasionally do harm. 'Do you think there are any apartments without ghosts?' asks Yau echoing de Certau's sentiment about the omnipresence of haunted places: '[t]here is no place that is not haunted by many different spirits hidden there in silence' (1988, p. 108). With no proper grave to rest in, the dead are likely to stay apartment-bound. It is no coincidence that the apartments in Hong Kong have been named 'coffin homes' after all.

The titular *jiangshi* is an interesting metaphor to consider. As the dead were meant to be buried near the place of their origin, in the old days, those who died away from home were supposed to be transported back by a rather unusual method that involved a Taoist priest known as a 'corpse walker' who used his magic to make the dead hop all the way to their destination. The hopping dead were known as the *jiangshi* and their awkward movements were generally explained as caused by the stiffening of their limbs, although it could also be suggested that the *yin* energy of the ghost was being repelled by the *yin* energy of the earth causing the corpse to hop. Whatever the reason, the *jiangshi* is essentially a creature that exists disconnected from land. The same can actually be said about an apartment—a housing unit that is always separated from the soil it stands on. The ownership of a house, or a grave for that matter, comes with the right of land, the ownership of an apartment does not. With the majority of apartments being rented rather than bought, human presence in them tends to be transient. In apartments, we are all ghosts.

Loneliness and Isolation: Invisible Lives

The opening shots of Byeong-ki Ahn's *Apt.* (2006) show the night view of a high-rise estate in Seoul from a driving car. Dark concrete slabs of the housing blocks loom ominously against the black sky, each looking exactly like the next. The shaky blurry shots are superimposed on top of one another creating a 'ghosted' image—a hallucinatory mirage of darkness and light. Despite the movement, the landscape appears completely lifeless and alien. Separated from the cityscape, the estate looks lonely and isolated. The aggregation of housing blocks evokes the Freudian uncanny—the homely and familiar, yet hidden from sight and withheld from others as eerie and arousing fear. Having finally reached the titular apartment we are not surprised to find it haunted, but a bigger message of the film is that in a neoliberal city, like Seoul, we are all ghosts.

Since the 1960s, Seoul has undergone several redevelopments aimed to transform it into a competitive global city. Corporate conglomerates were given the task to regenerate the urban environment through the practices of creative destruction and reconstruction. Iain Watson argues that the current neoliberal urbanism of Seoul is the result of planned administrative urbicide—a murder of the city (2013, p. 310)—a form of coordinated political violence where the destruction of buildings eradicates a possibility of creating meaningful communities (Coward 2009, p. 14). In the neoliberal city, real estate is seen primarily as an investment opportunity. As the main purpose of the apartment building is to generate profit, the apartment itself becomes first and foremost an asset, or a commodity that can be rented out or sold rather than a home.

The story of *Apt.* is set in an apartment complex. A lonely career-woman, Seo-jin spends her days watching other people's lives through her window. She notices a disturbing coincidence: at 9.56 p.m. all lights in the building mysteriously go out and the next day one of her neighbours is found dead. Although the deaths continue nobody seems to be alarmed, apart from Seo-jin, who tries to warn her neighbours but is told to mind her own business. The only person willing to talk to her is a wheelchair-bound young woman, Yoo-yeon, who lives in an apartment across from Seo-jin's and just like her spends most of her days staring through the window. Eventually, we learn that Yoo-yeon is a ghost exacting revenge for the abuse she suffered at the hands of her neighbours when still alive. But the uncanny similarity between Yoo-yeon and Seo-jin shifts our attention to the theme of urban isolation and 'living ghosts'—people who 'already in their lifetime, resemble dispossessed ghosts in that they are ignored and considered expendable' (Peeren 2014, p. 14).

Seo-jin spends most of her time alone. Early in the film, she is approached by a woman in a red coat who throws herself under a train pulling her along and pleading—'I don't want to die alone'. Seo-jin manages to get away but the words continue to haunt her. After the incident she becomes withdrawn and afraid to leave her flat. Unlike Yoo-yeon, whose visible disability marks her as different but also turns her into an object of pity, Seo-jin's internal paralysis remains hidden, partially because it is a condition shared by everyone around her. Talking with Seo-jin, Yoo-yeon recalls that she used to spend her days alone, 'hoping someone would acknowledge her … I didn't exist, even if I were there'. Ironically, death seems to imbue her with agency since as a ghost she becomes significantly more active and visible, transformed from a quiet young girl into a vengeful 'woman in red'. In the scene of Yoo-yeon's suicide, the blood from her open veins colours her white nightgown red, altering her appearance. This symbolically links her with the woman in the red coat we saw throwing herself under the train before, and matches a common representation of vengeful ghosts of suicides across East Asia. But the red dress also stands in opposition to the dull pastel colours Yoo-yeon used to wear when alive, just like her former passivity is contrasted by her assertive actions as a ghost. As a ghost, Yoo-yeon is capable of possessing people, she can exact her revenge, and eventually claims the life of Seo-jin—not because of a grudge but because she no longer wants to be lonely. Isolated apartment living is unbearable even to a ghost. Seo-jin may have escaped one woman in red only to be claimed by another.

Sociologists agree that the living environment can contribute to social isolation. As early as in 1938, Louis Wirth argued that sociological consequences of urban living include 'the substitution of secondary for primary contacts, the weakening of bonds of kinship, and the declining social significance of the family, the disappearance of the neighborhood, and the undermining of the traditional basis of social solidarity' (pp. 20–21), and that the majority of social interactions in the city are 'impersonal, superficial, transitory, and segmental' (p. 12). Inhabitants of the neoliberal city, additionally find themselves 'embedded within a highly uncertain geo-economic environment, characterized by monetary instability, speculative movements of financial capital, global location strategies by major transnational corporations and intensifying interlocal competition' (Peck et al. 2009, p. 57) which contribute to the dissolution of human bonds. Whether rented on short-lease contracts or bought as an investment, apartments are essentially unhomely (*unheimlich*) and evoke a sense of temporariness. Their tenants rarely form meaningful relationships with neighbours or even 'personalise' their space. People come and go. Locked inside their apartments they are alienated from human contact and are lonely.

The apartment estates depicted in Asian horror movies are mostly in low-to-medium income range, built to accommodate as many people as possible and make maximum profit from every cubic centimetre of available space. These are not the luxury condos with lush gardens, gyms, restaurants and entertainment complexes built to comply with the eco-friendly sustainable community models twenty-first-century cities strive to meet. The estates featured in Asian horror films typically appear to be sterile blocks of concrete dotted with a multitude of small windows, implying a large number of apartments, or they are almost windowless, fringed by an external hallway with a row of identical doors. Public spaces have mostly been replaced by car parks, although we can occasionally catch a glimpse of a deserted communal swimming pool, or a playground with an obligatory sandbox and a few lonely swings. Trees and greenery have been reduced to a bare minimum. While not completely brutalist in architecture, the environment seems hostile, or at best indifferent and adds to the mildly apocalyptic mood of these films (Image 2).

The loneliness of apartment ghosts mirrors the loneliness of humans. Both appear to be a by-product of living in the city, particularly cities like Seoul that have been redeveloped to serve the interests of the global capital rather than its people. John T. Cacioppo and William Patrick argue that '[i]n many parts of the world, older societies are rushing to embrace the American commodity culture and the casual disregard for social bonds that gave rise to the exurb's anomie' (2008, p. 254) and for many of

Image 2 The apartment complex from *Apt.* (dir. Byeong-ki Ahn, 2006) and its ghost

them this change happens too fast. Rapid urbanisation creates '[a] landscape built for disconnection' (p. 256) and fosters urban isolation. One interesting side-effect of social isolation, however, is people's focus on parasocial connections, including those with 'anthropomorphised supernatural agents', such as gods, angels or ghosts (pp. 257–258). Apparently in the city, we need ghosts as much as they need us.

Collateral Damage: Ghosts of a Dream

Although Asia is known for its cities, not all of them follow the same pattern of development. The megacities of Southeast Asia have established themselves as 'primate cities [that] dominate national economies and contain a disproportionately large percentage of a country's urban population' (McGregor 2008, p. 133). Their ongoing growth is driven by the constant migration of people from rural to urban areas in search of employment, or improved access to health services and education (p. 137). Since the 1960s, when Thailand began its transformation from agrarian to market-driven industrial economy, Bangkok has benefited disproportionally from the process and reinforced its economic dominance over the rest of the country. In 1960, Bangkok's regional per capita GDP was over five times higher than that of the Thai northeast (Askew 2002, p. 57) and by 2000, it produced more than a third of the country's GDP (UNFPA 2017, p. 103). By the early 1980s, Bangkok's population was twice that of the second largest Thai city, and by 2000, it amounted to more than half of the country's urban population (p. 103). This has greatly exacerbated regional inequalities and widened the symbolic rural–urban divide, where 'village and city ... stand as icons of distinctive kinds of people, places, and ways of life' (Mills 2012, p. 89). Mary Beth Mills calls Bangkok 'the privileged center of national sophistication' (p. 89) ideologically positioned as the centre of wealth and progress, with the rest of the country reduced to the role of the supplier of a flexible and inexpensive migrant labour force. Such uneven development, in turn, drives subsequent migration, which expands the city and deepens the divide even further.

The most obvious negative effect of such model is the fact that it makes the city particularly vulnerable to the challenges faced by the country's economy, since the city has become virtually equivalent with the country. In 1997, after more than a decade of unprecedented double-digit growth, Thailand became the first victim of the Asian Financial Crisis. The crisis hit Bangkok hard, exposing 'decades of unsustainable, shallow and dependent economic development' (Askew 2002, p. 91), resulting in massive lay-offs and closures of factories. The unemployment rate increased from 2 per cent of the total labour force in 1996 to 5% in 1998 and 5.3% in 1999 and suicide rates sky-rocketed to their highest peak in Thai history (Lotrakul 2006, p. 92). One of the most observable effects of the crisis were the shells of unfinished construction projects, some of which continue to mar Bangkok's skyline twenty years later. One such project that abruptly came to a halt was the Sathorn Unique. This 49-floor structure overlooking the Chaophraya river was planned to house over 600 residential units. The construction stopped when the building was 80–90% finished. At 185 metres, it remains one of the tallest abandoned skyscrapers in the world and a chilling monument to the blunders of past governments (Image 3).

News about the building becoming a magnet for suicides (the latest one in 2014) and rumours that it was built on the grounds of a former graveyard have earned it the ill-famed title of the 'Ghost Tower' (Kang 2017). The abandoned building and its ghosts identify the city as a Gothic space and a site of collective and personal trauma. While Bangkok rose to power thanks to its attachment to royal, religious and governmental institutions, its millennial incarnation has been built upon the dreams of migrant labourers who dedicate their lives to it in exchange for a promise of success. Most of them will remain invisible and exploited in their vulnerable professions as factory and construction workers, street vendors or domestic servants, and many will die in the city without ever achieving their goals. In Thai popular imagination, Bangkok is a city filled with

Image 3 The Sathorn Unique Tower in Bangkok (unfinished)

ghosts. Ironically, however, Bangkok's ghosts are awarded more attention than the people who walk among them—the 'living ghosts' of the city, and the ghosts of their dreams.

In 2017, Sathorn Unique became the location of a Thai horror film *The Promise* by Sophon Sakdaphisit that returns to the events of 1997, when the Thai economy collapsed. The plot focuses on the story of Ib and Boum, two teenage daughters of bankrupt property developers who make a suicide pact to die at the unfinished tower. Ib goes through with the plan, while Boum breaks her promise and grows up to become a developer herself. Seventeen years later, Ib's ghost returns to claim the life of Boum's daughter and stop her from completing the construction. While the film follows a conventional betrayal–revenge scenario, it is difficult not to look beyond this storyline to the events unfolding in the background. The titular promise, after all, encompasses much more than just an agreement between two friends. When the 1997 financial crisis put an end to the promise of transforming Thailand into one of the Asian Tiger Economies, it also decimated many dreams of a large part of the Thai population.

The film makes the point that the actions of the parents affect the lives of their children, just as the actions of the government affect the livelihood of the nation's citizens. The opening introduces a montage of home videos documenting the lives of Ib and Boum before the crisis. The girls are shown horse-riding, ice-skating, in a private pool, in a Mercedes, owning a pedigree dog, and playing the piano inside a luxurious mansion filled with an array of expensive-looking collectibles, a display of ivory tusks and other signifiers of their families' wealth. They seem oblivious of their privilege and unable to grasp its sudden loss. The film does not dwell on the tragedy of the adults choosing to introduce its aftermath through their daughters. Ib's face gets progressively more bruised, implying that her father has become abusive. Ib's words, however, suggest that she despises her father not as much for beating her but rather for his inability to take responsibility for the bankruptcy—he has a gun but no courage to kill himself. Boum resents her parents' decision to move into a small apartment and sell off their possessions. Having witnessed her father's public mental breakdown, she refuses to be embarrassed any further. She calls Ib and asks her to bring her father's gun to the tower. The girls make a pact to die together but after Boum witnesses Ib's suicide using her father's gun, she reneges on her promise and flees the scene.

Ib and Boum represent the dynamics that drive the primate city. Ib's ghost spends twenty years loitering around an unfinished building. She is a vengeful spirit with a score to settle but just like the ghost-tower she inhabits, she is also the ghost of a dream of all those who come to the city hoping to earn a better tomorrow for their children. Boum, with her focus on wealth and unethical conduct, plays the role of the ghost-maker. As a teenager she triggered her friend's death. Twenty years later, she has progressed to evading corruption charges, manipulating business partners and bullying her workers. As a result of her conduct, her company goes bankrupt and her daughter ends up in a coma. By the end of the movie, Boum has been reduced to yet another 'living ghost', forced to move into the same apartment she used to hate as a teenager. Ib's ghost has retained control over the tower, contributed to Boum's company's collapse and forced Boum to experience the same anguish the girls' parents experienced in 1997. The film's narrative evades closure—we do not know what happens to any of the characters. However, twenty years after its creation, the Sathorn Unique stands as a dark blot on Bangkok's shiny skyscape—an urban legend that haunts us with spectral visions of unrealised futures.

Conclusion: Ghosts in the City

City ghosts are a reminder that the uncanny is born out of the familiar. A haunted apartment shares none of the uniqueness of a haunted house—it offers no visible thresholds that should not be broken. Apartment ghosts are disconnected from the random succession of tenants that pass through the building, many of whom have already been reduced to the living ghosts themselves by their economic status, social isolation or health problems. The scenarios of Asian urban ghost films of the Anthropocene differ. Some films engage with local economic issues leading to the state-imposed segregation and dispossession of the urban poor, others comment on global migration trends and the destructive processes of aligning the development of cities with the needs of financial capital rather than their people. Occasionally, we encounter more apocalyptic visions of the post-human future where everything that remains is a ghost. The apartment—a housing unit rather than a home—contains an ecosystem that seems to exist independently of the land and, by extension, of the entire planet. It isolates its inhabitants from the world but also makes the world oblivious of their existence. The ghosts roaming apartment blocks successfully terrorise their random tenants but the real source of fear in these movies is that we have reached the point when nobody cares about that because we are all on the verge of disappearing anyway.

Notes

1. For a more detailed discussion of the significance of the Chinese ghost story model in the region see my chapter on 'Strange Ghosts: Asian Reconfigurations of the Chinese Ghost Story' in *The Routledge Handbook to the Ghost Story*, eds. S. Brewster and L. Thurston. London: Routledge, 2017, pp. 260–269.
2. The term is most commonly translated as the hopping vampire, thanks to the popularisation of the *jiangshi* by Rickie Lau's 1985 film, *Mr. Vampire*, although the creature does not consume blood. The *jiangshi* in *Rigor Mortis*, however, is significantly more destructive than its predecessors.

References

Apt. (*Apateu*). 2006. Byeong-ki Ahn. South Korea: Showbox.

Askew, M. 2002. *Bangkok: Place, Practice, Representation*. London: Routledge.

Berling, J. 1992. Death and Afterlife in Chinese Religion. In *Death and Afterlife: Perspectives of World Religions*, ed. H. Obayashi, 181–192. New York: Praeger.

Blundy, R., and H. Davis. 2017. Why Dying in Hong Kong Is Getting More Complicated… and Expensive. *South China Morning Post*, 15 July. https://www.scmp.com/news/hong-kong/education-community/article/2102544/why-dying-hong-kong-getting-more-complicated-and.

Cacioppo, J.T., and W. Patrick. 2008. *Loneliness: Human Nature and the Need for Social Connection*. New York: W. W. Norton.

Coward, M. 2009. *Urbicide: The Politics of Urban Destruction*. London: Routledge.

Curtis, B. 2008. *Dark Places: The Haunted House in Film*. London: Reaktion Books.

De Certeau, M. 1988. Walking in the City. In *The Practice of Everyday Life*, ed. M. De Certeau and trans. Steven Rendall. Berkeley: University of California Press.

Fischer, S., and F. Herschend. 2010. The Urban Mind Is the Normalcy of Urbanity. In *The Urban Mind: Cultural and Environmental Dynamics*, ed. J.J. Sinclair et al., 195–220. Uppsala: Uppsala University.

Gerow, A. 2002. The Empty Return: Circularity and Repetition in Recent Japanese Horror Films. *Minikomi* 64 (2): 19–24.

Gordon, A.F. 2008. *Ghostly Matters: Haunting and the Sociological Imagination*. Minneapolis: University of Minnesota Press.

Hoover, M., and L. Stokes. 1998. A City on Fire: Hong Kong Cinema as Cultural Logic of Late Capitalism. *Asian Cinemas* 10 (1): 25–41.

Juranovszky, A. 2014. Trauma Reenactment in the Gothic Loop: A Study on Structures of Circularity in Gothic Fiction. *Inquiries Journal/Student Pulse* 6 (5): 1–4. Available at http://www.inquiriesjournal.com/a?id=898 [15 July 2019].

Kang, D. 2017. 'Ghost Tower' Haunts Bangkok 20 Years After Financial Crisis. *Khaosod English*, 1 March. Available at http://www.khaosodenglish.com/news/bangkok//03/01/ghost-tower-haunts-bangkok-20-years-financial-cri sis/ [24 May 2019].

Lim, J. 2005. *Between Gods and Ghosts*. Singapore: Marshall Cavendish Editions.

Lotrakul, M. 2006. Suicide in Thailand During the Period 1998–2003. *Psychiatry and Clinical Neurosciences* 60: 90–95.

McGregor, A. 2008. *Southeast Asian Development*. London: Routledge.

Mills, M.B. 2012. Thai Mobilities and Cultural Citizenship. *Critical Asian Studies* 44 (1): 85–112.

Peck, J., N. Theodore, and N. Brenner. 2009. Neoliberal Urbanism: Models, Moments, Mutations. *SAIS Review of International Affairs* 29 (1): 49–66.

Peeren, E. 2014. *The Spectral Metaphor: Living Ghosts and the Agency of Invisibility*. Houndmills: Palgrave Macmillan.

'Regulators Now Spooked by Ghost Stories.' 2008. *Reuters*, 14 February. Available at http://www.reuters.com/article/us-ghosts-idUSN1442888920080214?feedType=RSS&feedName=oddlyEnoughNews [15 July 2019].

Rigor Mortis (*Geung si*). 2013. [Film] Juno Mak. Hong Kong: Kudos Films.

Scott, J.L. 2007. *For Gods, Ghosts and Ancestors: The Chinese Tradition of Paper Offerings*. Hong Kong: Hong Kong University Press.

The Promise (*Phuean thi raluek*). 2017. Sophon Sakdaphisit. Thailand: GDH 559.

UNFPA. 2017. *Impact of Demographic Change in Thailand*. Bangkok: United Nation Population Fund. Available at https://thailand.unfpa.org/en/publications/impact-demographic-change-thailand-0 [25 May 2019].

Watson, I. 2013. (Re)Constricting a World City: Urbicide in Global Korea. *Globalizations* 10 (2): 209–325.

Wirth, L. 1938. Urbanism as a Way of Life. *American Journal of Sociology* 44 (1): 1–24. Available at http://www.jstor.org/stable/2768119 [25 May 2019].

Arundhati Roy's *The Ministry of Utmost Happiness* as New Urban Gothic

Molly Slavin

In June of 2017, Arundhati Roy published her long-anticipated second novel, *The Ministry of Utmost Happiness*. Though a complicated and nuanced text that spreads across the South Asian subcontinent, the novel can be understood, as Parul Sehgal writes in *The Atlantic*, into a 'hulking, sprawling story' with 'two main strands'. The first strand follows Anjum (born Aftab), who works to carve out a space for herself in the city of Delhi[1] in the twenty-first century. The other strand of the novel is concerned with Tilo, an architect turned somewhat unwilling Kashmiri independence activist, who splits her time between Delhi and the Kashmiri capital of Srinagar, with a little bit of time in her home state of Kerala thrown in. Throughout the narrative, the two women and their friends and acquaintances repurpose and reinvent traditional Gothic tropes in order to articulate a vision of hope, regeneration and transformation. *The Ministry of Utmost Happiness* uses the new urban Gothic sites of Delhi and Srinagar to translate colonial and Gothic conventions for productive, justice-oriented purposes in contemporary South Asian cities.

M. Slavin (✉)
Georgia Institute of Technology, Atlanta, GA, USA
e-mail: molly.slavin@lmc.gatech.edu

H.-G. Millette and R. Heholt (eds.),
The New Urban Gothic, Palgrave Gothic,
https://doi.org/10.1007/978-3-030-43777-0_11

The Ministry of Utmost Happiness is Arundhati Roy's first novel since 1997's Man Booker Prize-winning *The God of Small Things*. Though she published much nonfiction in the intervening years and was quite prolific as a political activist for various causes, *The Ministry of Utmost Happiness* was, when published, bound to draw comparisons to her well-known and much-loved novel of two decades prior. Though *The God of Small Things* is unquestionably a Gothic novel, set as it is among crumbling ruins in a mysterious rural area with taboos surrounding sex and family abounding, *The Ministry of Utmost Happiness*'s Gothic tropes may need a bit more elucidating before proceeding. Roy's portrayal of South Asian cities and their contemporary inequalities, combined with her focus on death and graveyards, political violence, colonial legacies and neocolonial realities, consciously plays on nineteenth-century colonial and Gothic tropes to deliver a twenty-first century urban Gothic novel, revised and adapted for the postcolonial Indian and Kashmiri context. Much like in *The God of Small Things*, which emphasizes the characters' many-layered relationships to the United Kingdom, the British Empire hangs like a spectre over *The Ministry of Utmost Happiness*, particularly the violent hangovers of the 1947 Partition of British India, which split the colony into the contemporary states of India and Pakistan without providing a specific plan for the now-contested region of Kashmir. By giving us 'a tale of two cities'–Delhi and Srinagar—Arundhati Roy asks the reader to consider the implications of how colonialism and its legacies persist in Gothic literary tropes in a contemporary South Asian urban setting to create a literary language suited for a new type of urban Gothic.

The Ministry of Utmost Happiness begins in the Muslim *mohallas*, or neighbourhoods, of the Indian capital city, Delhi, focusing in particular on a young child named Aftab. Aftab, categorized as a boy when born, evolves to inhabit a more nuanced gender identity during childhood. A doctor who describes himself as a 'sexologist' tells Aftab's parents that the child is not, 'medically speaking, a Hijra – a female trapped in a male body – although for practical purposes that word could be used' (p. 20). The concept of *hijra* does not precisely translate to the contemporary Western definition of transgender; for one thing, *hijra* is recognized as a third gender in India, as well as many other South Asian countries, and *hijras* often, as Gayatri Reddy says, 'renounce sexual desire and practice by undergoing a sacrificial emasculation', after which 'they are believed to be endowed with the power to confer fertility on newlyweds or newborn children' (p. 2). *Hijras* are often understood to play ritualistic roles in Indian society, which makes a figure like Aftab, later Anjum, who opts out of societally prescribed roles, all the more remarkable and complex.

In *Queer Others in Victorian Gothic*, Ardel Haefele-Thomas makes the case that 'the *hijra* can often be an unassimiliable figure; this remains true today' (2012, p. 46). Though, Haefele-Thomas continues, *hijras* can sometimes be called upon to fulfill specific societal roles, they 'are not terribly welcome at any gathering' (p. 46); they are treated as monstrosities, outcasts, Gothic nightmares. Though Thomas is speaking of a figure in Wilkie Collins's 1868 novel *The Moonstone*, the *hijra* is understood by many as a pariah figure in twentieth- and twenty-first-century Delhi as well. After Aftab begins dressing in women's clothing and renames herself 'Anjum', she moves out of her parents' house and into the Khwabgah, or 'House of Dreams', a ramshackle old house populated entirely by those who identify as *hijra*. The leader of the house, Kulsoom Bi, lectures Anjum on the history of the Khwabgah, saying, 'This house, this household, has an unbroken history that is as old as this broken city... These peeling walls, this leaking roof, this sunny courtyard – all this was once beautiful' (p. 53). Anjum takes up residence in the Gothic mansion and never speaks to her father again; her mother will only meet with her in secluded places and then only sporadically and secretly.

Anjum's residence might be the most tangibly Gothic part of *The Ministry of Utmost Happiness*'s Delhi, but Roy takes care to demonstrate that the city as a whole remains structured by other traditionally Gothic or historic elements. Because, as Lucie Armitt writes, 'the Gothic is a deeply sociopolitical mode of writing' (date 2012, p. 510), and as Victor Sage and Allan Lloyd Smith say, the Gothic 'is not merely a literary convention or a set of motifs: it is a language [...] which provides writers with the critical means of transferring an idea of the otherness of the past into the present' (1996, p. 1), the Gothic is the perfect mode for Roy to utilize in making her points about the colonial past irrupting into the present to take on new, dangerous forms in the twenty-first century. Though *The Ministry of Utmost Happiness* takes place long after the British have left India, traces and ghosts of imperialism still remain, especially with regards to unjust economic and social structures. As Ken Gelder has observed, 'the postcolonial Gothic can certainly activate the traumas that result from a character's experience of loss and displacement after colonialism' (p. 198), and *The Ministry of Utmost Happiness* engages with several of Delhi's colonial and postcolonial traumas. Delhi is, at times, expressly neocolonial in its structure, with the Indian economic elites seemingly mirroring their former British rulers and millions of surplus people being forced out of the city, 'nobody knew where to' (p. 102). 'People who can't afford to live in cities shouldn't come here', a Supreme Court justice shrugs at one point, with a Lieutenant Governor adding, 'Paris was a slimy area before 1870, when all the slums were removed' (p. 102). The novel is full of depictions of how the poor must struggle to survive in Delhi amidst continuous power cuts, food shortages and dilapidated or nonexistent housing; yet, 'the food shops were bursting with food. The bookshops were bursting with books. The shoe shops were bursting with shoes' (p. 103). Roy puts it bluntly when she writes, 'Except that in Delhi there was no war other than the usual one – the war of the rich against the poor' (p. 398). The outright comparison of contemporary Delhi to nineteenth-century Paris recalls other structures, once thought buried and forgotten; a colonial-type administration that is openly callous towards common citizenry, unfettered and unbridled capitalism with no social safety net, beggars being 'held in stockades' (p. 103) for the crime of asking for food. The Gothic and imperialism have a long history, as writers including Patrick Brantlinger and Tabish Khair[2] have argued; with the reappearance of a kind of economic colonialism without the British in Delhi, Roy is, in *The Ministry of Utmost Happiness*, resurrecting colonial Gothic tropes for a twenty-first-century urban setting, highlighting to her readers the connections between the colonial Gothic and the new urban Gothic.

In 1961's *The Wretched of the Earth*, Frantz Fanon famously drew a Manichean picture of the colonial city as divided into the '"native" sector' and the colonizer's domain. While, according to Fanon, the 'native' sector is 'a disreputable place inhabited by disreputable people', the colonizer's part of the city, 'is a sector built to last, all stone and steel. It's a sector of lights and paved roads, where the trash cans constantly overflow with strange and wonderful garbage, undreamed-of leftovers' (p. 4). This image of the bifurcated city, with its Gothic overtones of the interplay between dark and light, death and life, has persisted into the twenty-first century, allowing postcolonial writers to recast colonial tropes for the postcolonial age; as Justin D. Edwards notes, Gothic 'figures and tropes are repeated across time and space, but they also shift and change according to the specificities of historical contexts, cultural legacies, and ideologies' (p. 50). In *The Ministry of Utmost Happiness*, Arundhati Roy adapts Fanon's sharply demarcated urban space for the new millennium, writing of two of her characters, both friends of Anjum, who dwell in the 'native sector' but are temporarily sojourning in the wealthier part of the city, that they were 'in a part of the city they oughtn't to be in. No signs said so, because everything was a sign that any fool could read: the silence, the width of the roads, the height of the trees, the unpeopled pavements, the clipped hedges, the low white bungalows in which the Rulers lived' (p. 139). The two, 'in a part of the city they oughtn't to be in', threaten to destabilize the balance of power in Delhi; as Mary Douglas has written, 'the danger which is risked by boundary transgression is power. Those valuable margins and those attacking forces which threaten to destroy good order represent the powers inhering in the cosmos' (1966, p. 162). By their mere presence, and their transgression into the colonizers' part of town, they are Douglas's 'matter out of place' (p. 36); they are Gothic aberrants, sent to subvert the colonial equilibrium of the city. The two friends of Anjum's move into a 'residential colony' where 'the houses were all three and four storeys high' (p. 141), in order to look after a baby, Miss Jebeen the Second. This baby, with Anjum as one of her eventual guardians, will play a large role in continuing to activate the colonial and Gothic elements of their surrounding urban space in the novel; Anjum, by inhabiting the 'traditional' *hijra* role with respect to taking care of children, further solidifies the text's investment in utilizing Gothic tropes to overturn colonial urban patterns.

Roy's use of the word 'colony' indicates the deliberate patterning of the city: even 70 years after the British left, Delhi is still divided into residential colonies for the rich and 'slums and squatter settlements, in resettlement colonies and "unauthorized" colonies' for the poor (p. 102).[3] Moreover, capitalism is failing postcolonial Delhi: Roy writes that while 'the surplus children slept, dreaming of yellow 'dozers' (p. 105) that will clear out their homes, 'jet aircraft darted about like slow, whining comets. Some hovered, stacked ten deep over the smog-obscured Indira Gandhi International Airport, waiting to land' (p. 105). This is a city in which, while the rich get richer, a baby—here, Miss Jebeen the Second—must be left on the corner in Delhi by a woman who cannot care for her. In Roy's description of Delhi, we readers see that rich Indians, in some ways, have taken the place of the imperial British so that the deliberate segregationist planning of the city persists, reinventing colonial structures for an ostensibly postcolonial age. Anjum, eking out a subsistence living in her Gothic mansion in a city that recalls nineteenth-century colonial settlement patterns, is Haefele-Thomas's 'unassimiliable figure' (2012, p. 46) who acts as a bridge, able to '[transfer] an idea of the otherness of the past into the present' (Sage and Smith 1996, p. 1). Though Anjum watches television and rides in cars, she and her surrounding city are reminders that the past is, to appropriate William Faulkner's line, not dead; it's not even past.

In *The Ministry of Utmost Happiness*, Roy is not only focused on those who struggle to survive in India's capital, but is also deeply invested in exploring the relationship of Delhi to Srinagar, the capital of the disputed territory of Kashmir, and in articulating what the interplay between these two capital cities means for the (post)colonial situation of these Gothic twenty-first-century urban spaces. When Roy first takes readers to Srinagar, we are greeted with a city that has been 'locked down. Security was being put in place for the funeral procession for the people who had been killed over the weekend, which would rage through the streets the next morning. There were shoot-on-sight orders' (p. 222). The complete militaristic control over the city by the Indian army strongly recalls colonial methods, and the overwhelming focus on death is resonant of Gothic tropes and patterns. 'The sullen city was wide awake but feigning sleep', Roy tells us (p. 54).

> By mid-morning the curfew would be lifted and the security withdrawn to allow people to reclaim their city for a few hours. They would swarm out of their homes in their hundreds of thousands and march to the graveyard, unaware that even the outpouring of their grief and fury had become part of a strategic, military, management plan. (p. 234)

The intertwining of present-day colonialism (this time, the Indian army is the colonizing force) and the ensuing death and destruction points to an articulation of a new type of urban Gothic, one in which the colonialism is further doubled and folded; the former colony, India, is now the colonial power over its own urban underclass, as well as Kashmir, leading to Delhi's Gothic tropes being refracted and articulated across the border in Srinagar.

While in Srinagar, *The Ministry of Utmost Happiness* turns its attention to Tilo, a woman from Kerala[4] who becomes entangled with Musa, a Kashmiri freedom fighter whose daughter, Miss Jebeen the First, was killed in the violence enacted by the Indian military in Kashmir. Tilo eventually adopts a baby who was abandoned on a street in Delhi and names her Miss Jebeen the Second in Musa's daughter's honour, articulating a kind of resurrection or refusal of death in the midst of an anti-colonial uprising.[5] Roy is interested in the Kashmiri independence movement, having written about it in *Capitalism: A Ghost Story* (2014) and in the collection *Kashmir: The Case for Freedom* (2011), as well as in many news outlets and assorted essays, putting forth her view that India is the colonizing force, and its power is felt both in Delhi and in Srinagar. The colonialism that manifests in the twenty-first century comes to the forefront with actions like slum-clearing or the persecution of religious minorities. In classically Gothic language, things thought long-dead are in fact coming back to life in new ways, even as the ghost of the former colonial occupier is still haunting these spaces.

Though Fred Botting and Justin D. Edwards argue, when discussing the 'globalgothic',[6] that in the twenty-first century, 'it is not always easy to locate tangible forms of empire, particularly given their transnational nature' (p. 17), this is not always true, as Arundhati Roy labours to show us. The contemporary conflict in Kashmir is 'the biggest, bloodiest and also most obscure military occupation in the world [...] The killing fields of Kashmir dwarf those of Palestine and Tibet' (Mishra date, p. 1). The Indian occupation of Kashmir and the city of Srinagar is unquestionably a twenty-first-century colonialism, made all the more layered because of India's former history as a colony: as Angana P. Chatterji writes, 'In the administration of brutality, India, the former colony, has proven itself equal to its former colonial masters' (dates, p. 96). Kashmir, being 'the unfinished business of Partition' (Ali 2011, p. 29), is both a remnant of the British Empire that has persisted into the current day, as well as a permutation of a new, vicious form of colonialism, perpetrated by the Indian government.[7] In '*Azadi*: The Only Thing Kashmiris Want',[8] Roy writes that nothing 'can justify the continued military occupation of a nation and a people. No more than the old colonial argument about how the natives were not ready for freedom justified the colonial project' (pp. 70–71). 'The Indian military occupation of Kashmir', she continues, 'makes monsters of us all. It allows Hindu chauvinists to target and victimize Muslims in India by holding them hostage to the freedom struggle being waged by Muslims in Kashmir' (p. 71). This imagery—of the (Indian) colonizer as monster—is repeated throughout her depictions of the Srinagar–Delhi relationship in *The Ministry of Utmost Happiness,* and highlights the Gothic elements of these urban spaces and their relationship to each other. The contemporary Gothic offers us a 'ready-made language for the aesthetic and cultural politics of our times' (Sage and Smith 1996, pp. 3–4). Roy's new Urban Gothic marshals a series of tropes and figures to react against the oppressive systems of the twenty-first century.

Roy repurposes the Gothic in order to empower individuals, as well as full societies. Back in Delhi, after a few years' residence in the Gothic Khwabgah, Anjum gets swept up in a religious riot in Gujarat and is arrested and imprisoned in the immediate fallout. After imprisonment in a camp (where she was forced to live among the men), she comes back to Delhi and, due to traces of lingering trauma, cannot re-adjust to life in the Khwabgah. Eventually, she packs up her belongings and moves into a graveyard. It was, Roy tells us,

> an unprepossessing graveyard, run-down, not very big and used only occasionally. Its northern boundary abutted a government hospital and mortuary where the bodies of the city's vagrants and unclaimed dead were warehoused until the police decided how to dispose of them [...] If they were recognizably Muslim they were buried in unmarked graves that disappeared over time and contributed to the richness of the soil and the unusual lushness of the old trees. (p. 62)

Anjum's decision to live in a graveyard recalls several traditional Gothic tropes. Eve Kosofsky Sedgwick notes that 'the convention of live burial' (p. 20) is common in Gothic fiction; while Anjum is not buried under the ground, her decision to live in a place of burial is certainly resonant of spatial metaphors of burial and being trapped. Jerrold E. Hogle concurs, saying 'antiquated settings' such as 'graveyards' (2014, p. 4) are a continually recurring theme of the Gothic, and Anjum's initial banishment from her family and subsequent self-exile to a graveyard recalls Maggie Kilgour's note that 'the gothic is thus a nightmare vision of a modern world made up of detached individuals, which has dissolved into predatory and demonic relations which cannot be reconciled into a healthy social order' (1995, p. 12). Anjum seems, for a moment at least, to be a 'detached individual', cut off from any community; the reader would fairly assume that a *hijra* living alone without the protection of the Khwabgah or any kind of community might be vulnerable to 'predatory' incursions like the type Kilgour describes.

Yet, Anjum is able to reinvent her initially atomized existence, with its potential for danger, into a 'healthy social order'. Roy demonstrates for readers the potential of colonial Gothic tropes reinvented for a postcolonial audience, building off the initial Gothic vision of 'detached individual[ity]' to eventually articulate a justice-oriented vision of a community: the graveyard allows her to recover from her trauma, and she begins to build rooms around each individual grave, welcoming in other outcasts from the neighbourhood as well as 'down-and-out travelers' (p. 72). Anjum eventually names her graveyard guesthouse 'Jannat' or 'Paradise', which eventually 'become a hub for Hijras who, for one reason or another, had fallen out of, or been expelled from, the tightly administered grid of Hijra Gharanas' (p. 73). By building a new existence out of sorrow and opening her doors to create community, Anjum is able to turn the Gothic motif of death into a vibrant symbol of life, much as the buried bodies in the graveyard produce 'the richness of the soil and the unusual lushness of the old trees' (p. 62). What might here be seen as an allegory of the gender transitioning process—the killing of one gender to fully flourish in another—also works well as a reference to the idea of communal rebirth. Jacqueline Rose writes, 'To survive, we all have to be seen. A transsexual [sic] person merely brings that fact to the surface, exposing the latent violence lurking behind the banal truth of our dependency on other people. After all, if I can't exist without you, then you have, among other things, the power to kill me' (2016) Anjum, by forcing herself to be seen by others, refuses death and instead embraces life.

Graveyards persist as a trope and a connective tissue throughout *The Ministry of Utmost Happiness* that work to further develop Roy's Gothic lexicon. The urban spaces Roy presents us with are overwhelmed by death and decay, but graveyards act as a counter-intuitive way to bring what is seemingly dead back to life, in an inverse relationship to the Indian military bringing old-fashioned colonial occupation back to life in Kashmir (which in turn brings death).[9] In addition to Anjum making her home in a graveyard and eventually turning it into a guesthouse to welcome all manner of outsiders and outcasts, we are told that, 'As the war progressed in the Kashmir Valley, graveyards became as common as the multi-story parking lots that were springing up in the burgeoning cities in the plains' (p. 325), that because 'Death was everywhere. Death was everything, [...] Graveyards sprang up in parks and meadows, by streams and rivers, in fields and forest glades. Tombstones grew out of the ground like young children's teeth' (p. 320). Musa, the Kashmiri freedom fighter romantically entangled with Tilo, writes to his dead daughter, Miss Jebeen the First, 'In our Kashmir the dead will live forever; and the living are only dead people, pretending' (p. 349). The continued presence of the dead in both Kashmir and Delhi, and the overwhelming refusal of the dead to go away, whether their staying power manifests in the physical property of the graveyard or a more spectre-like understanding of metaphysical presence, reiterates to the reader that this Gothic tale is continually excavating what had long thought to be passed away. Though 'in those days the little city of Srinagar died with the light' (p. 354) with the military-imposed curfew setting in at sunset, each day, Srinagar rises from the dead, ready again to resist the neocolonialism of the Indian army.

The Ministry of Utmost Happiness closes with one final act of resistance, this time economic in nature. A resident of Anjum's graveyard guesthouse, Saddam Hussain,[10] wants to take the other residents on a special outing, which turns out to be a trip to a gleaming, Westernized mall. The residents all troop out to 'where the countryside was trying, quickly, clumsily, and tragically, to turn itself into a city' (p. 415)[11] and enter this paradise of capitalist consumerism. While the younger residents seem fairly at ease in the setting, most of the older members of the community look 'as though they had stepped through a portal into another cosmos' (p. 416). After wandering around and window-shopping for awhile, the group has lunch at a Nando's, a globalized chicken chain from South Africa, another former British colony. It is in the food court that Saddam tells his assembled audience that he has taken them on this trip to introduce them to his father. 'This is where he died', Saddam tells them. 'Right here. Where this building now stands… My father's spirit must be wandering here, trapped inside this place' (p. 417).

Saddam's father's grave, then, is a sterile, well-lit, sanitized mall. His village was cleared to make way for this temple of capitalist and Western-influenced consumerism, and he remains there today, acting as a reminder of how the past continues to influence the present. Anjum is deeply uneasy with this, saying 'The matter can't be left like this. Your father should have a proper funeral' (p. 418). The group decides to buy a shirt for him from one of the mall stores and bury it in the guesthouse graveyard as a proper memorial and burial.

On the day of the funeral, Tilo, who has taken up residence in the graveyard with Miss Jebeen the Second, announces that she would like her mother's ashes to be buried in the old graveyard as well, making the day a double funeral. The day then becomes doubly marked, by death and by celebration. Though the characters remain in Delhi, a segregated city indelibly imprinted with colonial legacies and neocolonial realities, and most have strong emotional or physical links to Srinagar, the little city that dies with the light but resurrects every morning, they are able to resist these oppressive systems in the unlikeliest of places: a graveyard. Miss Jebeen the Second, indeed, begins her life in this place, 'a place similar to, and yet a world apart from where, over eighteen years ago, her young ancestor Miss Jebeen the First had ended hers. In a graveyard' (p. 311). The assembled residents, refusing to allow Saddam Hussain's father to rest in a mall, allowing Miss Jebeen the Second to simultaneously begin a new life and act as a reminder of the colonialisms that killed her predecessor, create 'the ministry of utmost happiness' in a graveyard. By creating this sense of community and allowing things to thrive where they should be dead, the residents are able to resist colonial logics via Gothic tropes in an effort to reconcile social relationships into a 'healthy social order' (Kilgour p. 12), carving out a space for a new kind of Urban Gothic.

Notes

1. Though the official name of the city is 'New Delhi,' Roy refers to the city as 'Delhi' fairly consistently throughout, as most of the novel is set in the 'old city' or 'old Delhi'. I have chosen to similarly refer to the city as 'Delhi' to reflect Roy's authorial choices.
2. See Brantlinger's *Rule of Darkness, British Literature and Imperialism, 1830–1914* (1988) and Khair, *The Gothic, Postcolonialism, and Otherness: Ghosts From Elsewhere* (2009).
3. 'Slum' began life as a nineteenth-century British slang term, while 'colony' is derived from Latin and subsequently Middle English; both are not only English words, but British words that hearken back to particular time periods in British colonial history.
4. Tilo's life story sounds as though it has been taken from that of the family who is the focus of *The God of Small Things.* Tilo is first introduced as being the daughter of a Syrian Christian and a man from an Untouchable caste, which mirrors the love story between Ammu and Velutha in Roy's first novel. Her mother's last name is later revealed to be Ipe, which is the same surname as the family at the centre of *The God of Small Things*, and her brother (Tilo's uncle) went to Oxford, which is the same biography as Uncle Chacko in the first novel. Tilo's mother later starts her own school, which was Ammu's dream. Whether these details are meant to flesh out Ammu's story in more detail, provide an alternate narrative pathway for the character, or perform some other function is left unexplained.

5. We learn that Miss Jebeen the Second's mother was a rural Communist activist who gave up her baby after being raped by Indian soldiers, briefly drawing in another geographic space to emphasize another neocolonialist aspect of the Indian military.
6. Glennis Byron, in her edited volume *Globalgothic*, asks readers to consider texts that 'are gothic in that we recognise their use of specific tropes and conventions' but that 'are produced and read in new temporal, geographical and political contexts' (2013, pp. 3–4).
7. The British Partition of India, a haphazard and last-minute affair that split up British India into the dominions of India and Pakistan in the span of six weeks, did a particularly poor job of dealing with the disputed territory of Kashmir. Both Pakistan and India laid claim to the princely state, and the British left the subcontinent without a clear plan or direction for the majority-Muslim area. Mass migrations and large-scale transfers of populations followed, as very few people wished to be religious minorities in the new states of either Pakistan and India; as is so tragically often the case, these migrations were accompanied by horrific physical and sexual violence. Today, the region remains split: the Kashmir Valley, Jammu and Ladakh are administered by India, while Pakistan lays claim to Azad Kashmir and Gilgit-Baltistan, and China holds the regions of Aksai Chin and Trans-Karakoram. The region is still heavily militarized, and sporadic violence breaks out, especially along the border regions.
8. 'Azadi' roughly translates to 'freedom'.
9. In a related meditation on the undead, Roy tells us that once Anjum moves to the graveyard, she becomes obsessed with 'B-grade Hollywood vampire movies and watched the same ones over and over again. She couldn't understand the dialogue of course, but she understood the vampires reasonably well' (pp. 72–73).
10. Not his real name; notably, he adopted this alias after seeing a video of the Iraqi Saddam Hussein's hanging, telling Anjum he admired the man's steeliness in the face of death, that he wanted to be the kind of person who can 'do what I have to do and then, if I have to pay a price, I want to pay it like that' (p. 95).
11. The implication that this space was only available for a mall after being cleared of its previous residents is fairly clear.

References

Ali, T. 2011. The Story of Kashmir. In *Kashmir: The Case for Freedom*, ed. T. Ali, H. Bhatt, A.P. Chatterji, H. Khatun, P. Mishra, and A. Roy. London and New York: Verso.

Armitt, L. 2012. The Magical Realism of the Contemporary Gothic. In *A New Companion to the Gothic*, ed. D. Punter. Chichester: Wiley.

Botting, F., and J.D. Edwards. 2013. Theorising Globalgothic. In *Globalgothic*, ed. G. Byron. Manchester: Manchester University Press.

Brantlinger, P. 1990. *Rule of Darkness: British Literature and Imperialism, 1830–1914*. Ithaca: Cornell University Press.

Byron, G. 2013. Introduction. In *Globalgothic*, ed. G. Byron. Manchester: Manchester University Press.

Chatterji, A.P. 2011. The Militarized Zone. In *Kashmir: The Case for Freedom*, ed. T. Ali, H. Bhatt, A.P. Chatterji, H. Khatun, P. Mishra, and A. Roy. London and New York: Verso.

Douglas, M. 1966. *Purity and Danger: An Analysis of Concepts of Pollution and Taboo*. London and New York: Routledge.

Edwards, J.D. 2013. She saw a Soucouyant: Locating the Globalgothic. In *Globalgothic*, ed. G. Byron. Manchester: Manchester University Press.

Fanon, F. 1961. *The Wretched of the Earth*, trans. Richard Philcox, 2004. New York: Grove Press.

Gelder, K. 2014. The Postcolonial Gothic. In *The Cambridge Companion to the Modern Gothic*, ed. J.E. Hogle. Cambridge: Cambridge University Press.

Haefele-Thomas, A. 2012. *Queer Others in Victorian Gothic: Transgressing Monstrosity*. Cardiff: University of Wales Press.

Hogle, J.E. 2014. Introduction: Modernity and the Proliferation of the Gothic. In *The Cambridge Companion to the Modern Gothic*, ed. J.E. Hogle. Cambridge: Cambridge University Press.

Khair, T. 2009. *The Gothic, Postcolonialism, and Otherness: Ghosts from Elsewhere*. Basingstoke and New York: Palgrave Macmillan.

Kilgour, M. 1995. *The Rise of the Gothic Novel*. London and New York: Routledge.

Mishra, P. 2011. Introduction. In *Kashmir: The Case for Freedom*, ed. T. Ali, H. Bhatt, A.P. Chatterji, H. Khatun, P. Mishra, and A. Roy. London and New York: Verso.

Reddy, G. 2005. *With Respect to Sex: Negotiating Hijra Identity in South India*. Chicago: University of Chicago Press.

Rose, J. 2016. Who Do You Think You Are? *London Review of Books* 38 (9) (5 May): 3–13. Available from https://www.lrb.co.uk/v38/n09/jacqueline-rose/who-do-you-think-you-are. Accessed 17 September 2018.

Roy, A. 1997. *The God of Small Things*. New York: Random House.

———. 2011. *Azadi*: The Only Thing Kashmiris Want. In *Kashmir: The Case for Freedom*, ed. T. Ali, H. Bhatt, A.P. Chatterji, H. Khatun, P. Mishra, and A. Roy. London and New York: Verso.

———. 2014. *Capitalism: A Ghost Story*. Chicago: Haymarket Books.

———. 2017. *The Ministry of Utmost Happiness*. New York: Alfred A. Knopf.

Rudd, A. 2010. *Postcolonial Gothic Fictions from the Caribbean, Canada, Australia, and New Zealand*. Cardiff: University of Wales Press.
Sage, V., and A.L. Smith. 1996. Introduction. In *Modern Gothic: A Reader*, ed. V. Sage and A.L. Smith. Manchester: Manchester University Press.
Sedgwick, E.K. 1986. *The Coherence of Gothic Conventions*. New York and London: Metheun.
Sehgal, P. 2017. Arundhati Roy's Fascinating Mess. *The Atlantic*, July/August. Available from https://www.theatlantic.com/magazine/archive/2017/07/arundhati-roys-fascinating-mess/528684/. Accessed 17 September 2018.

Urban Gothic: Singapore

Gina Wisker

Like infrared night-time cameras capturing the wanderings of the endangered pangolin around university premises and gardens, the urban Gothic expressions of Singapore reveal rich layerings of lives and histories, hidden from the intense, managed traffic flow. The soaring buildings, planned spaces and fast roads hide the past, yet its imprints are revealed by the Gothic imaginations of the writers who know what was there, what is hidden, what is not revealed to the tourist or investor. These giant buildings are new builds but carry the history of their locations and those who inhabited them—they are rebuilt on the grounds of others, sometimes on burial grounds or on the sites of wartime devastations. Partially revealed histories of imperialism, invasion, alternative medicines and lifestyles jostle alongside soaring glitzy capitalism. Everywhere there is the suppression and the explosion of life, rich, fertile, natural, whether it be the frogs and insects which invade your shoes and chocolate boxes, or reminders of what is hidden, dead, but not gone, the haunting presences and howling silences of ghosts of the bustling past whose dark secrets have been nearly

G. Wisker (✉)
University of Bath, Bath, UK
e-mail: gw647@bath.ac.uk

H.-G. Millette and R. Heholt (eds.),
The New Urban Gothic, Palgrave Gothic,
https://doi.org/10.1007/978-3-030-43777-0_12

erased, but exist parallel to the present, imprinted on the spaces, places, buildings and in the living histories of those who inhabit the island city.

I was a teenager in Singapore, as the high rises, malls and managed streets, started to spread across the kampongs and the cemeteries, isolating the botanical garden. But with the secretive, free movement of teenagers and the camouflage of youth it was possible to see culturally inflected spaces, liberating, improving or dark liminal spaces in the overlay of streets and byways mapping the potential and actual boundary crossings, with their transgendered prostitutes, opium dens and death houses. There were unspoilt tiny tropical islands just off the coast, houses on stilts in the brown limpid water on the coast, mynah birds on the sitting room window ledge, and just the one big black snake, embarrassed at being seen casually moving across the lawn. Most of what I experienced, what I saw, is neither mentioned nor acknowledged in public, political discourses, newspapers or TV news. Though it was somewhat idealised in exhibitions, creating vignettes of a migrated and invaded past with managed vegetation, flora and fauna and a narrative of improvement and modernisation. Nor were the darker sides of the back room experiences of the big houses and the new blocks of flats ever mentioned, but Catherine Lim, Sandi Tan, Sharlene Teo, Liz Williams and Fiona Cheong write of them: Gothic spaces and hidden histories, the liminality of a Singaporian urban Gothic and the lives within it.

Singapore, soaring with skyscrapers, link roads and metros crisscrossing this rich island nation of constant growth, is fertile and furtive, a consumerist capital at the east-west crossroads—which made it wealthy in the first place, also made it a crossroads of cultures. Here, histories are erased and marketised, a clean, safe, presence both advertised and enforced. As a city, its buildings, constantly renewed and reshaped, grand and dilapidated, hold histories which are reconnected/reconstructed publicly in museums, and less publicly in the private documents and on the bodies of those who have been silenced and interred within its past. Some of the darker secrets are those of indentured labourers, of marginalised people, of suppressed and sidelined women and of the poorer, more silenced cultural groups, of Malays and Tamils. Mixing these rich dark pasts with superstition and religion burgeons like tropical growth into the dark flowering of folk horror and ghost stories. There are many books of short stories showing Singapore rich with ghosts, from local writers, visitors or those drawn to stay there. Some of these cross-genre boundaries, combine across crime fiction, noir, the supernatural and Gothic horror.

Opening up an urban Gothic reading of Singapore, or more accurately texts on Singapore, it is useful to consider the insights of *Globalgothic* (Byron 2013), since Singapore is a city of global trade and cultural richness, and of tropical Gothic (Edwards and Guardini Vasconcelos 2016) since representations of Singapore can emphasise a sense of defamiliarisation related to its tropical heat and profuse (sometimes undead) life. *Postcolonial Gothic* (Punter 2000; Khair and Hoglund 2012) exposes a fractured homage to its imperial and colonial pasts of intersected cultures, trading and invasions, and Gothic tourism (McEvoy 2016), highlights its promise of a mix of world class shopping and staged memories of a culturally diverse, rich past, as well as dangers, just behind the carapace of managed differences. Exploration starts with Sandi Tan's work on pre-war, wartime and post-war Singapore in *The Black Isle* (2012). Tan's work shows Singapore full of historical ghosts jostling to be noticed and to have their stories told and their versions of imperialistic Singapore invaded, gendered power relations revealed.

In *Real Cities: Modernity, Space and the Phantasmagorias of City Life* (2005) using the work of Simmel and referencing Ian Sinclair's psycho-geography of the city in his walks in London, Steve Pile says 'I have sought to open up a field of analysis that is capable of taking seriously the imaginative, fantastic, emotional, the phantasmagoric aspects of city life' (p. 3) He is not alone in his interest in haunted cities. In their call for papers of a special issue of *Societies* (Gilloch 2013) the editors ask questions which are fundamental to postcolonial urban Gothic in particular because of the emphasis on whose memories and perspectives are prioritised, as they underlie thoughts here about Singapore urban Gothic, since it is such a multicultural history with a complex colonial and imperial past, and several cultural pasts erased by more dominant versions. 'How is the past etched into the very physical fabric of the modern cityscape? … Whose past is celebrated, whose memories are preserved? Are cities always and everywhere haunted by their pasts?' Their emphasis on amnesia is important since Gothic writers refuse to allow such blanking and silence, instead bringing into view the ghosts and other haunting presences imprinted on ostensibly brand new, or constantly renewed, places.

Particular views of the non-linearity of time underlie appreciation of ghosts and hauntings in cities. Singapore is our subject, but the ghost cities of China (such as Ordos, new, deserted, built to live in, only visited by commuters) emphasise an emptiness at the heart of an artifice of bustle, which speaks to the hollow core so often commented on. Tamara Wagner (2008) characterises Singapore Gothic as 'an urban Gothic of the nauseatingly mundane, of death-in-life literalized in the concealed spaces of anonymous apartment blocks' (2008, p. 58). Distressed urban Gothic spaces are an embodiment of a vacated set of human values.

Psychogeography (Debord 1955) is also useful in urban Gothic to explore and reveal both the everyday experiences for Singapore's urban complexities and its visual layout of cultural stratifications expressed in its streets: the central business districts; the reclaimed nature areas; the revived and rejuvenated historical sites; the hidden transgender cruising streets and bars; the buildings and areas which remain or are newly hidden from sight. It also helps reveal the hauntings of the past of this rich island nation, this sprawling yet manicured, contained, predominantly urban space, and the darker sides of such an intense capitalist focus on achievement and educational success, business success and the visible show of success through consumption. The Gothic reveals the dark side of excess: consumption, dedication to stratification and hierarchisation, social and cultural layering, new build investment, educational achievement, and its waste. Debord notes: 'the arrangement of the elements of the urban setting, in close relation with the sensations they provoke' (online no page), the effects of environment 'on the emotions and behaviour of individuals. The adjective *psychogeographical*, retaining a rather pleasing vagueness, can thus be applied to the findings arrived at by this type of investigation, to their influence on human feelings, and even more generally to any situation or conduct that seems to reflect the same spirit of discovery' (online n.p.).

The reorganisation and re-routings of urban spaces are for Debord to do with modernity and the attraction of abundance. This focus on ever-changing streets and spaces, emphasises a seeking after abundance, which in Singapore's case, emphasises abundance from consumer capitalism related to the desire to create effects of wealth, happiness, success. Physically, this involves foregrounding the beauty of huge malls and apartment blocks, a cleaned-up, tree-lined river, which once smelled foul; riverside shops, where once working go-downs and chop houses stood. The changes, visually, are to enchant business people and tourists, a pleasant and ostentatious indication of twenty-first century success and wealth. Latterly, in a rediscovery of creativity, Singapore is reclaiming some of its old buildings, reminding itself of its artistic heritage, making intentional artistic links with new contemporary art displays and buildings and an investment in art and dance. Debord's seeking an expression of abundance is in a continuum with hidden history, what lies behind the abundance, underneath and to one side of the tourist and business picture. Abundance has its dark Other in poverty and neglect. Both sides, both interpretations, are played out in the visible streets, the piers, malls, houses, hotels and hawker stalls. Singapore is exposed in its back streets and hidden spaces. Its lurking ghosts of the silenced past and embarrassing secrets are a parallel to its richness. Psychogeography conjures up the hidden histories and ghostly figures of those erased by development and success, writ large in the streets and buildings of the city, which is the island, rich with its criss-crossed history of invasion, imperialism, exchange, markets and cultural diversity. Its debilitating heat is held off in frozen malls, its layers of dubious past and present, lacquered over. Its darker side streets a proxy for the darker practices shunted off the main highways away from businessmen and tourists.

Tamara Wagner identifies 'engagement with different fictional conventions of negotiating an urban preternatural that renders the contemporary Singaporean urban ghost story so intriguing' (2008, p. 47). Building on legacies of Victorian fiction she argues that its

> urban Gothic importantly helps to untangle such flows of influence. It unearths the anxieties of a society at once repulsed and fascinated by the growing speed of its own change. Hence, the fictional representations of the haunted sites of exhumed graves, bulldozed shrines, and the high-rise block in all its anonymous sterility … [T]he Gothic of Singapore deals with concerns about development projects' … 'submerged or newly conceptualised heritage' … [and] suppressed aspects of turbulent histories, both on a national and an individual level. (2008, p. 47)

The many ghosts are of a repressed past newly disturbed. She sees Gothic scenarios as playing out a struggle with the past, present and future, and that this is part of a more extensive cultural re-formation, but emphasises also the 'traditional spookiness of liminal spaces of a fast-moving city', seeing them as 'remaking of the urban preternatural', concerned with issues of multiculturalism and hydbridisation (2008,

p. 48). There are other insidious repressions of course, and language is one of them. A significant part of the historical British empire, Singapore has its English language fiction rooted in that of Britain, leading to a homogenisation of the actually multi-ethnic Malaya/Malaysia and Singapore. It can be argued that literature and its expression of experience became, as it did in the Caribbean, a way of transforming, changing local experience and expression into something very British (often more British). English may seem a kind of neutral language for communication purposes, but it is actually far from neutral, since language derives from and perpetuates world views and value systems. Singapore may be silent about the Japanese invasion and the domination of a colonial past, about its hidden squalor and the oppression of minorities and, in every context, the silencing and oppression of women. Urban geography and construction help with this effacing but urban Gothic reveals it.

Hidden Histories: Sandi Tan, *The Black Isle* (2012)

Sandi Tan, Columbia University film graduate and film critic for *The Straits Times*, now living in Pasadena, is probably better known for her recently (October 2018) released film *Shirkers* (on Netflix), which she and a friend made in the 1990s when she was 18 and doing her A levels in Singapore. George, the American teacher who supported the girls and their friends in making the film, apparently ran off with 70 reels of it at the end of the summer, leaving a gap of over 20 years before his wife shipped them to Sandi. In her indie film she revisited her old self, combining a voice-over from the Sandi of the twenty-first century with the girl from the late 1990s (Montague 2018). In one interview she describes the growing heap of arriving film spools as 'an upright coffin' in her room. *Shirkers* is like a haunting of her past, and urban Singapore is its location. It is clear that 'Sandi Tan's past is teeming with ghosts' (Vankin 2012).

According to ghost seer Ling in Sandi Tan's novel *The Black Isle*, rich, pristine, fast moving, cosmopolitan Singapore, both island and island city, is 'dirty' with ghosts: 'The Isle shouldn't have to stay dirty. Ghosts don't have to live forever' (Tan 2012, p. 387). She sees them everywhere, reminding her of dark secrets, national and local violence and loss. *The Black Isle* focuses on Singapore as an island nation and a city invaded by the Japanese, bombed flat and brutally occupied while those previous imperial masters, the British, who were thought to help prevent the attacks and mitigate against the terrible damage, are seen wildly partying, in denial of responsibility. Relationships are overturned and undermined, and spaces both affect and reflect this. Singapore, damaged, rubble strewn, is crowded with the rapidly increasing numbers of new ghosts jostling for space and place, confused in their transitional/transitioning state. The traditions and the underpinnings of the ways of life and its hierarchies are similarly disturbed, destroyed and reconfigured in the upheaval of the invasion and the wrangling for power under Japanese rule and the capitalist development in the years after the war.

The Black Isle begins with an old Ling being followed then finally interviewed about her life with ghosts. The older woman retells the history of her younger self; and in this Tan mirrors the formula of *Shirkers*. The novel focuses on the years before, during and after the Second World War and the invasion of Singapore by the Japanese. Twins Li and Ling travel with their feckless father from Shanghai and their parents' broken marriage to Singapore, then Malaya (Malaysia today), living briefly on a rubber plantation, where their father fails to manage things, and they have to do most of the work. Imperialism enters with the owner of the farm, who is more feckless and demanding than their father but who is despatched one night by a Pontianak, a woman who died in childbirth who turns into a devouring ghost figure normally preying on other pregnant women, but here resorting to the over privileged slob.

Singapore is at a crossroads of imperial trade but described as both solid and with a tenuous hold on reality and security, floating in the sea through which trade and wars have been its history as a crossroads. It seems both colonial and solid:

> the Black Isle's place was as solid as Gibraltar, the British East India Company had owned and operated the Island since the early 1800 s, running it as a city-state, much like Florence or Venice at their peak, albeit with Anglo-Saxon stoicism rather than with Latin bravura. (p. 61)

Changing masters through history, its location also places it in a bustling but liminal space, a cosmopolitan trade route on the edge of a continent, fixed but fluid:

> Far from being the centre of things, however, the Island – as it was called by the locals – sat like a dull guest in the northwestern corner of the Archipelago, itself an unpromising clutch of crumbs with only the pull of the earth holding it in place. This arrangement seemed entirely provisional,

> and you got the sense from looking at the map that even the Isle might someday wear thin gravity's welcome and simply float away. (p. 61)

Opals are known to be bringers of bad luck and Tan describes Singapore as 'the shiny opal in the Empire's Far Eastern Crown' (p. 61). Her interpretation of the island and the city is that it is rich, ostentatious, but ultimately dangerously vulnerable and flawed. On Spring Street, Ling sees a tiger sauntering past terrified Englishmen, one of whom fails to kill it with a shot. Each part of the city is Gothic, with Tanglewood, the 'Black River, which snakes its way through D'Almeida place' (p. 80), blood-sucking mosquitoes, layers of backstreet houses behind more elegant fronts, and a history of pride in trade based on piracy. Li operates in the underworld, downtown, while Ling is brought to stay with the elderly wife of Mr Wee to protect her from thieving ghosts, which she does. Ling's great talent is her ability to live and perceive in the liminal space between the living and the dead. She experiences houses, rooms and streets in a parallel. There are ghosts everywhere in this pre-war city and Ling can commune with them. She is persuaded to use her gift to do more than protect Mrs Wee from theft. She agrees to work with Issa, the tough driver and guardian who is learning to talk with the dead at the Chinese cemetery and has almost finished a psychic spiritual course to become one who can truly commune with them. She strikes up a relationship with Wee's son Daniel, which leads to an engagement. Ling's survival and achievement are based on her spiritual insight and her adaptation to consumer capital contexts is reflected in her ability to enter the backstreets and downtown shops as easily as the big house owned by the Wees. Ling, like the ghosts she sees, can cross class and financial boundaries.

Singapore is a place of go-downs, backstreets and some grand houses, but when the Japanese invasion takes place, all is chaos as planes scream across the sky and there is no real cover in the streets, alleyways and shops. The invasion is like a terrible tsunami:

> High above, piercing the fog and the grey clouds, were flying metal crosses, appearing one by one until they formed a V. As the planes passed over our heads, the red circles on the base of their wings stared down upon us like blood-filled eyes. (pp. 224–225)

Hiding in a shop, Ling comes across twins who have lost their mother—she leaves, tries to help, but on return realises they are dead, adding to the ghosts in the city.

This Gothicised terror is bombing their streets destroying boundaries, lives, buildings and futures, while at another end of town the British are in an upstairs room partying and drinking, hurling bottles out of the windows that smash in the streets. The charade of imperialism's protection collapses in the moments of invasion, destroying people's safety and lives, turning the city to confusing rubble. Like one of Baudelaire's urban flaneurs (1869) but on a bike, Ling/Cassandra (after the soothsayer) can negotiate the terrain and comment on this loss and lack:

> I cycled on; the reek of alcohol smacked me as I approached the Cricket Club, a Georgian pile at the Padang's north end. Shattered glass glittered across its concrete front steps, not from any enemy bombardment. Weeping, ruddy-faced sots were gathered at the upstairs windows letting bottles of gin slip from their hands and smash to smithereens on the ground below. Inside a piano pounded out 'We'll meet again' accompanied by an off-key chorus. The Union Jack meanwhile, hung clammily on the flagstaff by the main door – it had never seemed more irrelevant or forlorn. (p. 236)

Teeming with ghosts in spaces interrupted by and created by the war, post-war Singapore is a Gothic-infused city of parallel worlds and histories. Crimes and losses are ever present and often demand restitution and different opportunities and curtailments are represented in different places. Ling/Cassandra negotiates all these spaces and their ghosts from her own liminality. Her skill of being able to see and speak to individual ghosts in crowds of angry and thwarted revenants is lucrative, as she is called to narrate their stories, ghost stories, as the city begins to be rebuilt. Survival and making a living for Ling/Cassandra depends on her abilities to interrelate the practices of the living with those of the dead and through her cunning, she transits from a vulnerable, poor, immigrant to wealthy mistress. With no fixed home of her own, she ironically communes with the spirits of the old homes that are earmarked for new buildings and homes for the rich.

The terror, violence, imprisonment and brutality, which the Japanese invasion brought, was silenced by rebuilding after the war. But as Sandi Tan shows, the ghosts of that war are ever present in the rebuilt streets and in the corners of rich restaurants, richer houses, reminders of that covered over past, that silenced history, and the whitewashed, glitzy, soaring skyscrapers of the re-routed streets. Ling occupies a fluid and transitional position between the invaded landscape, the new capitalist world and the encroaching spirit world.

She is a survivor who navigates the complexities of time, space, living and dead. In new luxurious apartments she brokers between the living and the dead, making deals over space and its use. On the edge of the island, the Wees' mansion is replaced with a public beach. Daniel's close friend, Oxford-educated Kenneth Khee, architect of post-war versions of Singapore, and Ling's lover knows no-one will want to party with disturbed and angry dead ancestors so he hides them in plain sight. His ambitious duplicity makes him fill this play space with Japanese lighting, making it difficult to see the hordes of ghosts hanging around the living holiday-makers. But Ling/Cassandra sees them. Ignoring or circumventing these dead citizens is always a mistake. With Kenneth Khee in charge, the first apartment blocks, which housed local communities, are either bought out or burnt out (arson), to make way for expensive new housing. Rich, new flats are built over the bulldozed, burned out rubble of the largely informal settlement of Redhill. However, as their homes are razed, multitudes of ghosts of the displaced dead build up their anger with Khee and his property development plans. Reference to boundaries, entrenched areas, entrapped spaces and death render the post-war Singaporean landscape one of urban Gothic horror:

> The dead were the ones who wailed. I tried to avoid them but they were all around me. Mothers shrieking for their sons, sons shrieking for their mothers. (p. 412)

Ling/Cassandra reports on this travesty, the displacing of people to the outskirts or the backstreets so that the rich can have luxurious apartments. Added to this is the development of a metro, which necessitates the digging up of cemeteries and leads to angry responses from long-time residents now displaced ghosts. Khee angers so many living and dead citizens that it is unsurprising when, visiting to attempt to placate them, he vanishes into the mound amid the clamouring of ghostly hands. He bought, developed and then evicted both the living and the established ghosts from previously secure homes so they take revenge as they can—pulling him into their own Forbidden Hill of the dead.

The ghosts are always there, and the Gothic of the place lies in both this ghostly reminder and its histories, the composites of the reconstruction of streets and buildings, the replacement and new fronts to the old routes of canals and roads and the routing of new metro lines through the cemeteries of the long dead. In Sandi Tan's work and in the Facebook posts of contemporary Singaporean friends and activists, it is clear there is protest against this erasure, and that the ghosts are not pleased.

Readers and audiences might be more attuned to Tan's exposé of the Gothic version of urban Singapore after the success of contemporary Singaporian lifestyles in the 2018 film *Crazy Rich Asians* (2018), which highlights the glitzy and superficial side of the super-rich.

Crazy Rich Asians (2018)

The film *Crazy Rich Asians* (2018) is in many ways a romantic comedy of the kind Margaret Atwood might produce, for example in *Lady Oracle* (1976), where the artifice and deception around romantic love leading to weddings and marriage (and/or deceptions and painful break ups) are laid open as delusive fantasy. Only a little Gothic, there are no readings of buildings and ghostings in the film, its first moral deficiency remains indulgence, in all its forms. Super-rich Singapore-born, Oxford-educated Nick Young woos Rachel Choo in New York, where they are equals in terms of intellect, jobs and lifestyles, but when he takes her home to Singapore for the wedding of close friends, the grotesque excesses of the super-rich lifestyles, and the tyranny and meanness of family and friendship groups, rivalries and power games turns the world through Rachel's eyes into an alienating prison of excess, of conspicuous consumption and spite. Dynastic control and rejection mingle amidst the visual excess of the soaring artificial cityscapes high above the glittering, shifting shapes of the city, while partygoers in clothes costing thousands, drink themselves stupid then pamper themselves with fashion statement clothing and beauty treatments that fight the tides of age. The Stag Party is a manic disco on an old ship to which the men are taken by helicopter, and from which they escape, and the hen party is a riot of shrieking sisterly spite. Rachel endures a dislocated Cinderella experience. She has won the prince of Singapore, and everyone is a mean, skinny, over beautified (non-ugly) 'sister', determined she will know she should not be and is not welcome at the ball of the rich. Normal friendships are defamiliarised. To emphasise her outsider status, the Sisters camouflage their immense jealousy only temporarily with artificial friendship before daubing her bed with what she describes as the bloodied taunts of a serial killer. Like the crazed party-goers in Edgar Allen Poe's 'The Masque of the Red Death' (1842), this group of socialites, are so rich and removed from reality that they can make her life hell with impunity. The matriarchal family into which she has naively entered, ensures she knows she is there on false pretences—a usurper, not even a Choo of note. Rachel's first view of the grounds and house remind of the reader of 'the second wife' glimpsing Manderley in Daphne du Maurier's *Rebecca* (1938) for the first time. She is dangerously out of place and struggles to step over their threshold of the glittering, expensive mansion with its profuse, excess of flowers. Although the old grandmother seems initially delicate and friendly, admiring Rachel, she then rejects her as utterly unworthy. The mother, like the queen in Snow White, is the woman of the house. She refuses half the food like a medieval king (who would have their own taster but expect poison). All potential partners for Nick will be rejected—not quite poisoned with an apple or, as in 'Sleeping Beauty', pricked with a poisoned needle, but threatened and demeaned by the matriarch. The mother is an arrogant controller of wealth whose matriarchal rights were conferred on her when her forefathers swept in years ago having held high office and then capitalised from this with land ownership and construction stretching across South East Asia.

Dangerous to this young woman Rachel, the family scenario and urban setting are a mixture of Gothic fairytales, each shot through with glittering clothing, food and ornate furniture costing millions. The urban Gothic of *Crazy Rich Asians* warns of shallowness and spite lying at the heart of excess. Luckily or necessarily the film finally sheds its Gothic exposure of hollowness amid the tropical growth of excess flowers and glittering buildings and returns to a Cinderella fantasy with a happy ending, although it is not certain where Rachel and Nick might live. The darkness revealed in this Gothic RomCom is the hollowness of relationships in the excessive urban space of Singapore. Rachel's defamiliarisation on entering this rich life, the meanness meted out to her, the falseness—water with rose petals flooding the church for the wedding (a marriage of deception)—make this an urban Gothic tale.

In his introduction to work on urban Gothic cities and spaces, Steve Pile, using Walter Benjamin, notes 'the significance of dreams, magic, vampires and ghosts for the emotional work of city life' (2005, p. 3), and when talking of Singapore in his vampire city chapter he describes its 'red lips, and sharp white teeth: sex and death' (2005, p. 112) as part of the city's fabric.

> Singapore's modernity and its urbanism is ghostly, haunted, just as it is magical and vampiric, Singapore is not alone in this – such experiences with ghosts expose the ghostly figures that parade through the phantasmagoria of modern city life, the hauntedness of modernity itself. (Pile 2005, p. 135)

In rich glitzy cities social acceptance and recognition are essential for survival and success, so in *Crazy Rich Asians*, Rachel's Chinese–American single earner family background almost entirely prevents any chance of that survival, let alone a romantic love ending with Nick, particularly when the spiteful rich bitches warn her off with a blood red threat.

Conclusion

Historically, Singapore sent its less publicly acceptable side offshore, down the back alleys, in its death houses or opium islands, hidden from view. The history of invasion, change, pretence and hypocrisy is accompanied and represented by the ghosts everywhere, which accompany and bewail events lurking in the corners of every house, present on every street, a historical reminder of the darker moments of Singapore's nineteenth-century past, the subject of Sandi Tan's *The Black Isle*. The swift cutting and re-cutting of motorways and roads, the massive building enterprises, give Singapore a sense of constant, unstoppable growth, change and movement. It's hard to keep your bearings, but beneath and beyond the foundations of the new buildings and the scoring of the new roads are the previous cartographical versions, and the ghosts.

There is not much room for the other ghosts of Singapore, island nation and sprawling city. Lush, heaving with life, the older Singapore bursts into organic development whenever it can—the tropical climate favouring rampant growth of flowers, tendrils, trees, insects and birds, but the cityscape forces concrete to cover much of the natural land, and the kampongs and hillside graveyards are razed for more buildings. So even the spirits of Singapore, as well as the restless, ever growing creatures, can feel themselves castrated by concrete, and suffocated or surrounded by high rises and concrete barriers. Singapore's urban Gothic is written into its constantly growing buildings, the growth managed by contractor's plans. Singapore is heaving with people, shoppers, commerce and tourists walking and taxi-ing through its streets. It is a crossroads of West and East, on the bottom of the Malaysian peninsula, in the centre of the silk and spice routes, makes it a space of transition and change. Its urban Gothic is one of metamorphosis, grand soaring concrete and glass constructions embodying human aspirations and achievement, and hiding the bases of power, money and the history of invasions, changing political allegiances, corruption and excesses, of stunted growths of hopes, dreams and plans. Singapore is a lively, beautiful hymn to capitalist consumerism and exchange, a hub for world trade and an urban Gothic hub of layers upon layers of lurking historical secrets expressed in its spatial configuration of soaring buildings and dark back alleys, lush tropical growth, and parallel existence of the rich and poor, the fixed and transient, the living and the dead. Some of Singapore's secrets are related to trade, to its immigrated Chinese, Malay, Indian and European cultural history, and to its vital, vulnerable position on the edge of Malaysia, which made it both an epicentre and a target for the silk and spice routes, for British and Japanese invasive imperialism, dubious investments and bird flu. Singapore is an island nation, a large urban sprawl with a shifting fabric of skyscrapers and fast living, a tiny 'teardrop island' (Tan 2012, p. 455), criss-crossed with racing motorways and concrete girders carefully overgrown with glorious bougainvillaea. Ghosts are ever present in the rapidly growing, changing, metamorphosing urban environment of Singapore; they are everywhere. As the city island has risen, spread, morphed, dissolved re-written and reconstructed itself, so it has often tried to forget and hide its ghosts, but this rich mix of Chinese, Malay, Tamil and European culture is not one that forgets easily or without consequence.

References

Atwood, M. 1976. *Lady Oracle*. Toronto: McClelland & Stewart.

Baudelaire, C. [1869] 1970. *Le Spleen de Paris*, trans. L. Varese. New York: New Directions Publishing.

Byron, G. (ed.). 2013. Introduction. In *GlobalGothic*. Manchester: Manchester University Press.

Debord, G. 1955. Introduction à une Critique de la Géographie Urbaine. *Les Lèvres Nues*, 6 September. http://library.nothingness.org/articles/SI/en/display/2. Accessed 11 August 2016.

Du Maurier, D. 1938. *Rebecca*. London: Victor Gollancz.

Edwards, J.D., and S. Guardini Vasconcelos (eds.). 2016. *Tropical Gothic in Literature and Culture of the Americas*. New York: Routledge.

Faucher, C. 2004. *As the Wind Blows and Dew Came Down: Ghost Stories and Collective Memory in Singapore*. New York: Routledge.
Frayling, C. 1991. *Vampyres: Genesis and Resurrection from Count Dracula to Vampirella*. London: Thames and Hudson.
Gilloch, G. (ed.). 2013. Special Issue 'Ghost-towns: Cityscapes, Memories and Critical Theory'. *Societies* 3 (3).
Joaquin, N. 1972. *Tropical Gothic*. Brisbane: University of Queensland Press.
Khair, T., and J. Hoglund (eds.). 2012. *Transnational and Postcolonial Vampires*. Basingstoke: Palgrave Macmillan.
McEvoy, E. 2016. *Gothic Tourism*. Basingstoke: Palgrave Macmillan.
Miéville, C. 2009. *The City and the City*. London: Macmillan.
Montague, T. 2018. I Had to Tell My Story: An Interview with Sandi Tan. https://www.rookiemag.com/2018/11/i-had-to-tell-my-story-an-interview-with-sandi-tan/.
Pile, S. 2005. *Real Cities: Modernity, Space and the Phantasmagorias of City Life*. London: Sage.
Poe, E.A. [1842] 1966. The Masque of the Red Death. In *Complete Stories and Poems of Edgar Allan Poe*. New York: Doubleday Dell Publishing Group, Inc.
Punter, D. 2000. *Postcolonial Imaginings: Fictions of a New World Order*. Edinburgh: Edinburgh University Press.
Tan, S. 2012. *The Black Isle*. New York: Grand Central Publishing.
Vankin, D. 2012. The Writing Life: Sandi Tan Switches Genres for "The Black Isle". *LA Times*. https://www.latimes.com/entertainment/la-ca-sandi-tan-20120805-story.html.
Wagner, T.S. 2008. Ghosts of a Demolished Cityscape: Gothic Experiments in Singaporean Fiction. In *Asian Gothic: Essays on Literature, Film and Anime*, ed. A. Hock Soon Ng. Jefferson, NC: McFarland.

Filmography

Crazy Rich Asians. 2018. dir. Jon. M. Chu.
On the Beach. 1959. dir. Stanley Kramer.
Shirkers. 2018. dir. Sandi Tan.

Suzhou River: 'On the [Haunted] Waterfront'

Annemarie Lopez

Lou Ye's film *Suzhou River* (2000) portrays an eerie vision of late twentieth-century Shanghai: a decaying city polluted by its industrial past, but also by a proliferation of untold stories. This chapter examines how the film draws on Hollywood Noir and Gothic tropes to evoke a spectral urban milieu. The principle negotiation of the film's narrative is that of the social and aesthetic malaise that has arisen from the rapid economic and environmental changes in turn-of-the-millennium China.

Lou Ye's *Suzhou River* is set in the late 1990s, when China decentralised its bankrupt state-owned studio system, allowing privatised film companies to take over production. Asian studies scholar Hongwei Lu argues that despite ongoing obstacles such as 'political censorship, financial restrictions, and Hollywood infiltration' the move allowed for more independent filmmaking and enabled filmmakers to tackle more controversial subject matter (Lu 2010, p. 116). *Suzhou River* was the first film made by one of China's first independent film production companies, Dream Factory, founded by Chinese filmmaker Lou Ye in association with Berlin-based German producer Philippe Bober. The film was well-received in Europe, winning prizes at international film festivals in

A. Lopez (✉)
London, UK

H.-G. Millette and R. Heholt (eds.),
The New Urban Gothic, Palgrave Gothic,
https://doi.org/10.1007/978-3-030-43777-0_13

Rotterdam and Paris, but was nevertheless banned from screenings in China (Lu 2010, p. 116). This is a multilayered film drawing on a range of cinematic and literary influences that showcase Lou's hybrid cultural fluency. Western film critics have pointed out its use of documentary-style hand-held camerawork, its references to classic Hollywood Noir and its specific narrative similarities to Hitchcock's *Vertigo* (Francke 2011; Rooney 2000). This essay considers how the film employs these cinematic influences as part of a wider aesthetic engagement with the Gothic; how spectrality is used to explore issues of representation in late twentieth-century China; and how this engagement expands our understanding of the Gothic beyond its more familiar Western contexts.

Suzhou River opens in darkness with a dialogue between an unseen man and woman (later identified as the narrator and his girlfriend Meimei). The woman asks: 'If I left you someday, would you pursue me?... Would you pursue me forever?' The man replies that he would, but the woman retorts that he is lying. The black screen dissolves into footage of Suzhou River, also known as Suzhou Creek or Wusong Creek, a major waterway in a former industrial zone of Shanghai. The river was highly polluted and in decline in the twentieth century, but was scheduled for regeneration in the late 1990s (Du 2014). Through the lens of the unseen videographer-narrator's camera, the viewer travels along the river, witnessing the murky waters, the empty and partially demolished warehouses along its banks and the gruelling, impoverished lives of the workers on its barges and demolition sites. The anonymous videographer describes how he likes to observe life along the waterway, a place full of stories. 'If you watch long enough, the river will show you everything', he says (Image 1).

Within its first few minutes, *Suzhou River* frames its narrative as a story about storytelling, and about the multitude of stories that take place along the river—the lives of the riverside workers we see, stories the narrator describes of a girl he once saw jumping into the river and the bodies of young lovers pulled from the water. The viewer is intentionally destabilised and disoriented and jump cuts between images add an element of visual confusion. We are not sure if the film we are about to see will be a love story, a documentary, a fantasy or if anything in this story can be believed. This elusive and unreliable quality becomes a recurring motif. The videographer tells us he *could* tell us a story about a mermaid but 'it would be a lie'. Later, however, he explains that he makes videos for people, but if people don't like what they see it's not his fault, because

Image 1 Suzhou River waterfront from *Suzhou River* (2000)

'my camera doesn't lie'. At this point the narrator seems to be suggesting a disjunction between what is shown and what we are told. The camera doesn't lie, but people and storytellers do. This motif is repeated with variations throughout the film. The videographer turns to his own story of how he was hired by a nightclub owner to film the club's resident mermaid act. He tells us that he didn't believe in mermaids until he saw Meimei (Zhou Xun) swimming in her fish tank, and then he fell in love. Of course, Meimei is not an actual mermaid, but appears as one in her performative mode, and so the narrator seems to be telling us that the appealing illusion made him fall for a fictional story. In turn, the narrator's story about two doomed lovers originates from a story his girlfriend Meimei told him, so while we witness their story on screen, we are never sure whose version we are seeing.

This recurring theme of a disjunction between visual evidence and fictional constructs is woven throughout the narrative threads of the film. In this way the film begins to draw us into its spectral world of ghosts and doubles. My use of 'spectral' draws on both its everyday usage as 'ghostly', and the hauntological notion of 'spectrality'. In his essay on hauntology, Colin Davies defines the spectral as that which replaces being and presence with the figure of the ghost 'which is neither present nor absent, neither dead nor alive' (Davis 2005, p. 373). Davis notes that hauntology encompasses a number of conceptions of the spectral and does not necessarily require a belief in actual ghosts, but suggests a 'deconstructive figure' that hovers between presence and absence, that calls into question the living present and makes 'established certainties vacillate' (Davis 2005, p. 376). Davis's argument draws on the notion of the spectral from Jacques Derrida's 1993 study *Spectres de Marx*, and its English translation (Derrida 1994). Davis notes that Derrida explored the idea of the 'phantom' in the work of psychoanalysts Nicolas Abraham and Maria Torok but took a different approach. Abraham and Torok's phantom is a figure that appears to conceal secrets and hidden memories of past traumas, which 'should be uncovered so that it can be dispelled' (Davis 2005, p. 379), but in Derrida's formulation of the spectral, the figure of the ghost does not seek to conceal this trauma so much as challenge us with it. Derrida's ghost is 'that which haunts like a ghost and, by way of this haunting, demands justice, or at least a response' (Peeren and Pilar Blanco 2013, p. 9). In *Souzhou River*, we see how the pairing of Moudan and Meimei (both played by Zhou Xun) performs a similar function, both haunting the videographer/narrator and Mardar and destabilising the viewer's experience of the events witnessed on screen.

An interest in spectrality extends beyond Western philosophy, as Chinese literature scholar Andrea Riemenschnitter observes, pointing to a wider 'spectral turn in global cultural production since the 1980s' (Riemenschnitter 2014a). Riemenschnitter argues that this spectral preoccupation is a response to a global cultural crisis provoked by the ever-encroaching impact of the neoliberal capitalist world. She also notes that the globalisation of these spectral themes has led to reciprocal encounters between Asian traditions of the supernatural, and the Gothic. For example, Riemenschnitter links the popular fox-headed 'eco-Gothic' spectre of the Na'vi from *Avatar* with the ancient Chinese tradition of *zhiguai* (tales of the strange or anomalous, often involving supernatural creatures, animal spirits and demons), and points out that modern versions of *zhiguai* appear in Asian cinema alongside localised versions of Western Gothic themes. Riemenschnitter's observations are highly constructive in considering *Suzhou River's* engagement with spectrality, ghosts, supernatural creatures and Western traditions of the Gothic.

As Gothic scholar Catherine Spooner points out, contemporary critics are moving towards 'an understanding of the Gothic as a set of discourses rather than simply as a genre' (Spooner 2007, p. 2). There is a constellation of ideas and concepts that might be considered Gothic, stretching back to the eighteenth-century novels of Horace Walpole and Ann Radcliffe, and echoed in later works by Mary Shelley, Bram Stoker, Charles Dickens and Robert Louis Stevenson, and as Dale Townshend argues, up to the dystopic tales of Cormac McCarthy's *The Road* and the film *World War Z*. (Townshend 2013, p. xxvii). Townshend acknowledges the elasticity of the Gothic over time, evolving from an eighteenth-century notion of 'the ancestral or the medieval past', to its modern literary-critical usage as 'a certain horrific and gruesome strain in fiction' (Townshend, xxxvi). However, Townshend argues that, 'at its most characteristic', the Gothic evokes 'the unbearable persistence of history, with a traumatic, painful and nightmarish past that, however deep our wishes, will not simply disappear' (Townshend 2013, pp. xxxviii–xxxix). This fixation on the past is often coupled, Townshend says, with an evocation of uncanny places, 'spaces in which we have been hurt and wounded, but to which we obsessively return' (Townsend 2013, p. xxxix). It is these two elements, the persistence of a troubling past and the depiction of uncanny space, that are key to understanding *Suzhou River*'s particular engagement with the urban Gothic.

Sara Wasson points out that the intense emotions of anguish and terror often aroused in eighteenth-century Gothic novels were accompanied by the evocation of claustrophobic and oppressive spaces such as the castles, monasteries, convents and dungeons of old Europe (Wasson 2010, pp. 2–3). However, the Gothic found its 'modern corollary' in the disorienting spaces of the modern nineteenth-century metropolis. Wasson suggests that the urban Gothic horrors of human violence and corruption were 'terrifying enough without necessarily needing a supernatural edge' (Wasson 2010, p. 3). She also argues that literary depictions of the city, and in particular London, reflected anxieties about class and capital, the dangers of urban anonymity that enabled sexual and criminal transgressions, and the modern city's diurnal rhythms, which came 'to represent interior psychic dramas' (Wasson, p. 3). In Lou Ye's *Suzhou River* an eerie urban atmosphere is evoked through scenes of gritty riverside life, crumbling apartments and smog-blurred skylines, alongside Noir-style images of neon-drenched streets, murky dive bar interiors and the greenish glow of Meimei's mermaid tank. The haunted and oppressive *mis en scene* is also enhanced by a plot that resonates with narrative echoes from the past.

Most notably, *Suzhou River* recalls Alfred Hitchcock's *Vertigo* (1958) (Rooney 2000; Francke 2011). In *Vertigo* a former policeman (Jimmy Stewart) becomes infatuated with a beautiful woman (Kim Novak) he has been hired to follow and protect. He fails to save her and when the woman falls to her death, Stewart's grief-stricken character becomes obsessed with a similar-looking shop girl (also played by Novak). He fixatedly tries to recapture his lost love through this woman, until he uncovers the truth that they are one and the same. In *Suzhou River* Mardar is hired to drive a young woman (Moudan) around town for her businessman father, falls in love with her, but betrays her by getting involved in a kidnapping plot with his criminal associates. When Moudan discovers the plot, she jumps from a bridge into the river, telling Mardar she will return as a mermaid. Mardar then becomes obsessed with Meimei who both resembles Moudan, and performs as a mermaid. Though played by the same actress, the film leaves us in doubt as to whether they are the same person. The similarities between the two stories are obvious, but also raise the question of how they differ. Ever since Laura Mulvey's influential essay 'Visual Pleasure and Narrative Cinema', Hitchcock's *Vertigo* has been read in psychoanalytic terms (Mulvey 1999). Mulvey wrote how the film was self-consciously and actively engaged in scopophilia or voyeurism. She argued that the use of subjective camera in *Vertigo* put the audience in the position of the male detective—as subject-viewer of the female object—laying bare the way in which cinema positions the female image as a threat to the male ego, who must be objectified ('castrated') to be enjoyed (Mulvey 1999, pp. 833, 840–842). *Suzhou River* also features a subjective camera—we see the story entirely through the eyes of the videographer and, as in *Vertigo*, *Souzhou River*'s narrative also ostensibly involves unravelling a mystery involving two similar women. Several theorists have argued, however, that Lou Ye's use of these *Vertigo* tropes is not mere mimicry, but something closer to what Riemenschnitter called 'a reciprocal encounter' between Western and Chinese aesthetics. This reciprocal encounter, as Riemenschnitter suggests, is not simply an adoption of Western themes, but involves placing them alongside Asian traditions and allowing them to engage with each other in 'a ghostly traffic between the living and the dead' (Riemenschnitter 2014a).

In *Hitchcock with a Chinese Face*, art historian Jerome Silbergeld argues that *Suzhou River*'s use of Western models such as Hitchcock, is not an example of cultural imperialism, or of 'colonising or commodifying the

Chinese subject for delivery to a foreign audience', but instead demonstrates Lou's skill in participating in the global culture while using it to 'address internal [local] concerns' (Silbergeld 2004, p. 6). Silbergeld points out that though there are some striking resemblances between *Vertigo* and *Suzhou River*—e.g. in both the films the 'reincarnated' heroine sits at a dressing table lit with green and pink neon from a sign outside her window—it is not a 'shameless rip-off' as one critic labelled it (Silbergeld 2004, p. 15). Instead, he argues that the resemblances recall the traditional Chinese artistic practice known as *fang*, or 'transformative imitation', a type of 'referential homage' in which the poet or painter works in the style of another but 'with a twist' (Silbergeld 2004, p. 13). The intention, says Silbergeld, is not slavish mimicry, but 'creative' borrowing, that often leads in 'directions quite different, or even contrary to the original work'. Hence, Silbergeld argues that whereas in Hitchcock's *Vertigo* the central narrative tensions stem from Freudian psychology, in *Suzhou River* the depiction of a pattern of 'betrayal, loss, pursuit of recovery', is less about crime-solving or decoding individual developmental psychology, than exploring 'collective neuroses' and a 'corrupting cynicism' in the social sphere (Silbergeld 2004, p. 18). Silbergeld argues that the application of Hitchockian Noir to Shanghai extends the dialogue about the semi-Western nature of the Chinese city, in the same way that the use of the mermaid image refers to Chinese hybridity (Image 2).

While part-fish part-human creatures can be found in Chinese folklore and are part of a tradition of images connecting women to water and ideas of sacrifice, Silbergeld argues that Lou is also clearly drawing on a Western tradition of mermaid stories. This is supported by Lou who admits to being introduced to Hans Christian Anderson's 'The Little Mermaid' by his parents and dismisses the idea that it fits into an easy east-west distinction.

> The whole idea [of the mermaid] is a Western import, like Coca-Cola and McDonald's - everything the Fifth Generation filmmakers refused to admit as Chinese... But how can you draw this line? As a child, I can remember my parents telling me Andersen's 'The Little Mermaid'. And when did my parents first hear it? It's impossible to say. This, I think, is very important with the filmmakers of my generation. We're not interested in being cultural immigration officials saying, this is Chinese, this is not Chinese. (Rosenbaum 2001)

Image 2 Hybrid human-fish creature from Chinese zhiguai classic, *Classic of the Mountains and Seas*, illustrated book, 17th century Japan, British Museum

Here Lou rejects the idea that symbols or stories can be purely Western any longer and stresses the importance of looking beyond the China-West division. For him, mermaids, like Hitchcock and Coca-Cola, have become symbols of the changing nature of Chinese culture, which has absorbed foreign influences and stories in order to fashion a new narrative for itself. This also recalls Riemenschnitter's earlier observation about the recent blending of Chinese traditions with localised versions of Western Gothic themes. Indeed, in upsetting simple paradigms or categories, the mermaid, half-human, half-fish, is perhaps a monstrous Gothic apparition of a new globalised hybridity.

Film scholar David H. Fleming has also explored the use of Western cinematic influences in *Souzhou River*. He notes that the film's moody tone and the generic trope of the unreliable urban hero who mixes with underworld types and becomes infatuated with an illusive night-club performer, identifies a 'distinct Noir mode at play' (Fleming 2014, p. 140). Fleming points to Deleuze's observations of how in Noir certain emotions (confusion, paranoia) become detached from individual characters and begin to 'expressionistically' pollute the whole film, evoking the post-war crisis in identity. *Suzhou River*, he argues, uses similar devices to evoke feelings of disorientation and disruption. In *Suzhou River*, however, Fleming suggests that characters 'emerge and move within memory and time rather than space' (Fleming 2014, p. 140). While the film draws on certain Noir tropes, Fleming convincingly argues that it is ultimately an exercise in aesthetico-political détournement aimed at 'modifying our understanding of the (cinematic) past' (Fleming 2014, p. 135). Furthermore, critic Gary G. Xu suggests that *Suzhou River*'s use of Hitchcock, and indeed its resemblance to other Western films including Scorsese's *Taxi Driver*, Godard's *Breathless*, and Kieslowski's *Double Life of Veronique* is part of its technique of misdirection (Xu 2007, p. 86). Xu suggests that Lou is in fact self-consciously deconstructing the idea of mimicry itself, blurring fact and fiction to highlight the 'Chinese postsocialist logic' of 'socialism with Chinese characteristics', which is in essence a submission to capitalism and globalisation in disguise (Xu 2007, p. 86). As such, Lou is critiquing the hypocrisies of this Chinese socialism in name only, by revealing the mechanics of imitation. Xu quotes Lou: 'When you actually enter the life of our generation, you will discover that the imitations by us who grew up in mimicry are actually ultimately original and creative'. Xu agrees with Silbergeld that *Suzhou River* is in fact, 'a global MacGuffin', constructing, 'under cross-cultural spectatorship' an apparent love story, while in fact attempting to show the reality of life on Suzhou River and the decaying reality of Shanghai (2007, p. 87). Taking on board these observations, *Suzhou River* also engages with the Gothic as a way of exploring the Shanghai cityscape and its relationship to its inhabitants. The film's use of *Vertigo*-esque characters, Noir plots and mermaid figures contribute to this Gothic engagement, serving as spectres of an unresolved past and repressed present. While the spectres of Meimei, Moudan and Mardar may be drawn from Hollywood tropes, they nevertheless appear alongside real people, in a real urban context, established in the documentary-style footage of Suzhou River, and serve to deconstruct that reality.

The film's depiction of Shanghai is centred around the river itself. It rarely strays far from the water, barges, dilapidated warehouses, neon-lit bars and cramped apartment buildings alongside its banks. The narrator tells us that if you watch long enough the 'river will show you everything', and indeed it has played a significant role throughout Shanghai's history. Descriptions of the once vast waterway appear in texts of the Tang Dynasty (618–907). In the nineteenth century it became a strategic site in the trade wars between China, Britain and America, and became a key conduit for international trade. In the twentieth century, however, the silted-up river had shrunk in size and become one of the most polluted waterways in the world. Awash with waste from the city's declining industries, it became known as 'the smelly river' (Du 2014). In the 1990s it was earmarked for an urban renewal project, which appeared to have just begun when *Suzhou River* was filmed (Grescoe 2016). An article in 2016 in *Shanghai Daily* claimed that by 2000 Suzhou Creek was 'once again clean, with fish and duckweed once again thriving in the river' (Zhang 2016). In choosing this setting for his tale of spectral romance and disillusionment, however, Lou's narrative suggests scepticism and disenchantment with this optimistic vision of turn-of-the-millennium Shanghai.

These specific local concerns also prompt a new set of aesthetic engagements with the Gothic. Carol Senf has argued that in its earliest form, the Gothic provided a counterbalance to the confident Enlightenment worldview in which science and rationalism swept away old superstitions, and instead 'recognized that the power of the past, the irrational, and the violent continue to hold sway in the world' (Senf 2014, pp. 31–32). Meanwhile, Robert Mighall has argued that the urban Gothic mode emerged from a specific critique of conditions in mid-nineteenth-century London experiencing the impact of industrialisation and the rise of the urban poor (Mighall 2007, pp. 54–72), in which the settings move from isolated castles and monasteries in old Europe, to the squalid modern city. Industrialised China (epitomised by The Great Leap Forward between 1980 and 1995), has superceded this industrialised urban Victorian era, with unprecedented levels of development, resource consumption and human displacement. Mighall argues that the urban Gothic landscape '[locates] terrors and mysteries in criminalized districts in the heart of the modern metropolis' (Mighall 1999, p. xxii). The decrepit buildings and labyrinthine streets of London were depicted as zones of decay, poverty, crime and confusion where characters become disoriented and prey to dark forces. In *Suzhou River*, concerns about pollution, poverty and displacement are evoked through images of the haunted, dilapidated waterfront. In the film, the spectres are not demons or murderers, but the lonely alienated figures on the bridges and barges, and the ghosts of Hollywood Noir and Western fairy tales. Yet they may be just as troubling for Chinese audiences, acting as reminders of the very real political and artistic censorship in China. Even the appearance of ghosts and Noir figures on screen is a sensitive issue in contemporary Chinese culture, where crime and supernatural themes have long been discouraged. Critic Li Zeng notes that Chinese cinema has long been dominated by social realism, and censors worry that graphic representations of violence and the dark tone of horror, might 'pollute young people's minds', while subjects such as ghosts and the supernatural 'promote superstition' (Zeng 2013, p. 110). Supernatural themes such as ghosts and mermaids have appeared in more recent Chinese films, but are always handled in oblique ways. Stephen Chow's *The Mermaid* (2016) became China's biggest ever box office success, but presented its mermaid in a narrative about rapacious capitalism and environmental concerns, rather than a Gothic or mythological context (Sala 2016). Meanwhile, Andrew Nette points out that crime fiction was banned in China after the communist revolution in 1949, and apart from a brief relaxation of censorship laws in the 1990s, crime stories have been strictly controlled as a topic for fiction and film as they are seen as undermining social harmony (Nette 2011). *Suzhou River* then must have been particularly problematic for Chinese authorities as it conjures images of both crime and supernatural themes and blurs the lines between the two genres. However, Lou is clearly aware his film is playing with the rules. When the narrator tells us that he could tell us about a mermaid, but he would be lying, he is obliquely reminding us of the difficulty and danger of telling stories in China. Yet by depicting a mermaid through the images of Moudan's mermaid doll, Meimei performing in her tank and sitting on the banks of the river combing her blond wig, Lou is able to conjure a mermaid spectre whose presence–absence in the cityscape calls into question these rules. Here the mermaid becomes a symbol not only of Chinese hybrid culture, but an image that tests what it is possible to see under the Chinese regime and what we can believe.

In this way *Suzhou River* not only expands the urban Gothic beyond its established historical and geographic context, but also recruits its supernatural tropes to deal with the question of what can be shown and said about contemporary Shanghai. The river is represented as an uncanny space, where the real people who live there are glimpsed through jump cuts, while their stories of poverty and struggle remain untold because they do not fit the official narrative. Instead their stories must be revisited elliptically, through Hollywood references and stories about doubles and ghosts. As critic Nicholas Royle has noted, the uncanny involves a 'feeling of uncertainty' particularly in regard to 'who one is and what one is experiencing' (Royle 2003, p. 1). Sigmund Freud's essay on the uncanny identifies situations which arouse this sensation of uncanniness, including uncertainty about whether something is living or dead (Freud 2001, p. 226), and the recurrence of certain faces, as in the figure of the double (Freud 2001, p. 234). While doubles and ghosts are a recurring motif in *Suzhou River*, their role is not so much to reveal the psychological underpinnings of its individual characters, but rather the nature of living in a society where the truth can never be spoken directly but must always be concealed within other stories.

Suzhou River was in the early stages of regeneration when Lou made his film, which shows workers demolishing the old buildings along the riverbank. Yet Lou also seems to be questioning the narrative of urban renewal which seeks to erase the past and offer the promise of a glorious future. Chinese state development policies have permitted the relentless destruction of communities and neighbourhoods, as Andrea Riemenschnitter has noted. Yet the response of Chinese artists, she argues, 'distances itself from the Western romantic and modernist appreciations of Gothic morbidity without denying their impact' (Riemenschnitter 2014a). Lou's depictions of the filthy river and the crumbling waterside warehouses do not indicate nostalgia, but question what will happen to the lives lived there. In a scene towards the end of the film where Moudan and Mardar are briefly reunited, they stand together facing towards the sunset, but with heads turned in different directions. Mardar looks down towards the river with a dejected expression, while Meimei looks up hopefully towards the Oriental Pearl Tower across the Huangpu River in the Pudong New Area. The radio and television tower, built in the mid-1990s, was China's tallest structure at the time and offers a vision of Shanghai's promised future. Earlier in the film, the tower provided the distant backdrop for Moudan's first plunge into the river from the bridge, and now it seems to foretell their doom. Ultimately, the tower is a cruel spectre, a symbol of a future that will never be fully realised and will come at a cost. Shortly after their reunion, the couple, drunk on bootlegged liquor from Moudan's father's store, drive into the river and drown. The tower becomes a symbol of a fatal fall from illusion, echoing the Spanish mission tower in Hitchcock's *Vertigo*.

After the couple's plunge into Suzhou River, all that remains is the spectre of the couple's lost love. Until that moment, we are told, Meimei didn't believe Mardar's story and didn't think that Moudan existed. Yet in death their love undermines Meimei's relationship with the narrator, conjuring an ideal she feels compelled to pursue, just as Mardar obsessively pursued his memory of Moudan. Meanwhile, the viewer is left to wonder whether this is merely a story the narrator has spun to explain the bodies of the drowned couple he saw pulled out of Suzhou River at the beginning of the film, and whose actual story will remain untold, like so many stories among the rubbish. Catherine Spooner has argued that in the new ecological manifestations of urban Gothic, the focus is on the tension between the city's surface, the 'civilized' developments of a city, and the intrusion of the persistent, undesirable, 'primitive, darker depths' (Spooner 2006, pp. 27–28). This kind of urban Gothic shifts its focus from the human psyche to a repressed dark side intrinsic to the city itself, and questions the very nature of contemporary urban progress. In *Suzhou River*, however, the polluted river also serves as a physical reminder of what cannot be said about this progress in China. Andrea Riemenschnitter describes a number of environmental incidents in 2013, when Beijing suffered its worst air pollution and at the same time, thousands of dead pigs and ducks floated down Shanghai's Huangpu River, which now serves as a source of water for residents. Despite assurances from the authorities that there were no health risks, netizens quickly began making jokes online, saying that in Beijing 'you only need to open the window for free smoking and in Shanghai there is free pork soup from the tap' (Riemenschnitter 2014b, p. 224). These ironic comments highlight the way in which Chinese citizens view their government's statements with scepticism, yet cannot directly criticise them. The truth must always be revealed obliquely, and Lou Ye's film suggests that this comes at a cost. In *Suzhou River* the result of unspoken truths is a type of psychic pollution, which manifests itself in the weary faces of many of the film's characters, and in the cynical and fatalistic tone of the narrator.

At the end of the film Meimei disappears, leaving the videographer a note that says: 'Find me if you love me'. But he does not look for her, and instead starts drinking. He tells us that this helps him briefly recapture his time with Meimei, as he conjures her spectral presence by replaying videos he recorded of her. His intoxication summons her ghosts and allows him to imagine an alternative reality in which they are still together, and 'the sun comes out and the river is clear and full of fish', thus equating his romantic fantasy with the regeneration of the river. The narrator tells us he won't run after Meimei, like Mardar did, because 'nothing lasts forever'. At this point the music changes, as it did in the opening of the film, from romantic strings to a menacing staccato rhythm of beating drums. The videographer tells us he will simply drink and close his eyes, 'waiting for the next story to start'. Mounted on a boat belching smoke, the camera travels along the murky Suzhou River until the screen turns black. Here the narrator tells us that he refuses to believe in illusions any more. He is ostensibly talking about love stories, but also about China's official narratives, propaganda about Shanghai's bright future and cities without pollution. Lou has in fact compared political disillusionment to a failed love story. Speaking about the lead up to and aftermath of the Tiananmen Square pro-democracy protests and massacre in 1989, the subject of his 2006 film, *Summer Palace*, Lou says: 'It was like falling in love. And then after 89, people felt like they had lost something, like they had broken up with a lover' (Watts 2006). In *Suzhou River*'s ambiguous final message, the narrator tells us that 'another story will start soon'. This time, he seems to imply, we are not obliged to believe it.

References

Davis, C. 2005. Hauntology, Spectres and Phantoms. *French Studies* 59 (3): 373–379.

Derrida, J. 1994. *Specters of Marx: The State of the Debt, the Work of Mourning, and the New International*, trans. P. Kamuf. New York: Routledge.

Du, Q. 2014. Free Flowing History. *Global Times*, April 7, viewed 5 October 2018. http://www.globaltimes.cn/content/852959.shtml.

Fleming, D.H. 2014. The Creative Evolution and Crystallisation of the 'Bastard Line': Drifting from the Rive Gauche into Suzhou River. *Journal of Chinese Cinemas* 8 (2): 135–147.

Francke, L. 2011. Mermaid's Tale: Film of the Month - Suzhou River. *Sight and Sound*, 20 December, viewed 8 January 2019. http://old.bfi.org.uk/sightandsound/review/520.

Freud, S. 2001. *An Infantile Neurosis and Other Works, Vol. 17, 1917–1919*. London: Vintage.

Grescoe, T. 2016. Shanghai's Suzhou Creek Cleans Up Its Act, November 26, viewed January 23, 2019. https://www.theguardian.com/travel/2016/nov/26/shanghai-suzhou-creek-clean-up-redevelopment.

Lu, H. 2010. Shanghai and Globalization Through the Lens of Film Noir: Lou Ye's 2000 Film, Suzhou River. *ASIANetwork Exchange: A Journal for Asian Studies in the Liberal Arts* 18 (1): 116–127.

Mighall, R. 1999. *A Geography of Victorian Gothic Fiction: Mapping History's Nightmares*. Oxford: Oxford University Press.

Mighall, R. 2007. Gothic Cities. In *The Routledge Companion to Gothic*, ed. C. Spooner and E. McEvoy, 54–72. New York: Routledge.

Mulvey, L. 1999. Visual Pleasure and Narrative Cinema. In *Film Theory and Criticism: Introductory Readings*, ed. L. Braudy and M. Cohen, 833–844. New York: Oxford University Press.

Nette, A. 2011. *Crime Fiction & Social Harmony in China*. The Wheeler Centre, Melbourne, viewed 23 January 2019. https://www.wheelercentre.com/news/4f42c3a19f0a.

Peeren, E., and M. del Pilar Blanco (eds.). 2013. *The Spectralities Reader: Ghosts and Haunting in Contemporary Cultural Theory*. London: Bloomsbury.

Riemenschnitter, A. 2014a. Beyond Gothic: Ye Si's Spectral Hong Kong and the Global Culture Crisis. *Journal of Modern Literature in Chinese* 12 (1), viewed 11 January 2019. http://www.ln.edu.hk/jmlc/content_f.html.

Riemenschnitter, A. 2014b. The Revindication of Environmental Subjectivity: Chinese Landscape Aesthetics between Crisis and Creativity. ARI Working Paper Series WPS 224, University of Zurich, viewed 17 January 2019. https://doi.org/10.5167/uzh-98340.

Rooney, D. 2000. Suzhou River. *Variety*, 16 February, viewed 8 January 2019. https://variety.com/2000/film/reviews/suzhou-river-1117778688/.

Rosenbaum, J. 2001. The World Is Watching: Suzhou River. *Chicago Reader*, 8 March, viewed 5 December 2018. https://www.chicagoreader.com/chicago/the-world-is-watching/Content?oid=904820.

Royle, N. 2003. *The Uncanny*. Manchester: Manchester University Press.

Sala, I.M. 2016. No Ghosts. No Gay Love Stories. No Nudity: Tales of Film-Making in China. *The Guardian*, 22 September, viewed 12 December. https://www.theguardian.com/film/2016/sep/22/tales-of-film-making-in-china-hollywood-hong-kong.

Senf, C. 2014. Why We Need the Gothic in a Technological World. In *Humanistic Perspectives in a Technological World*, ed. R. Utz, V.B. Johnson, and Travis Denton, 31–34. Atlanta: School of Literature, Media, and Communication, Georgia Institute of Technology.

Silbergeld, J. 2004. *Hitchcock with a Chinese Face: Cinematic Doubles, Oedipal Triangles and China's Moral Voice*. Seattle: University of Washington Press.

Spooner, C. 2006. *Contemporary Gothic*. London: Reaktion.

———. 2007. Introduction: Gothic in Contemporary Popular Culture. *Gothic Studies* 9 (1): 1–4.

Suzhou River. 2000. Dream Factory and Essential Filmproduktion GmbH, Berlin. Directed by Lou Ye.

Townshend, D. 2013. Introduction. In *The Gothic World*, ed. G. Byron and D. Townshend. Abingdon, Oxon: Routledge.

Wasson, S. 2010. *Urban Gothic of the Second World War: Dark London*. London: Palgrave Macmillan.

Watts, J. 2006. Camera Obscured. *The Guardian*, 9 September, viewed 23 January 2019. https://www.theguardian.com/commentisfree/2006/sep/09/comment.china.

Xu, G.G. 2007. *Sinascape: Contemporary Chinese Cinema*. Lanham, MD: Rowman & Littlefield Publishers.

Zeng, L. 2013. Ghostly Vengeance, Historical Trauma: The Lonely Ghost in the Dark Mansion (1989). *Journal of Chinese Cinemas* 7 (2): 109–121.

Zhang, Q. 2016. The Story of the Suzhou Creek. *Shanghai Daily*, 23 October, viewed 17 January 2019. https://archive.shine.cn/sunday/now-and-then/The-story-of-the-Suzhou-Creek/shdaily.shtml.

A Gothic Barcelona?: Carlos Ruiz Zafón's *The Cemetery of Forgotten Books* Series and Franco's Legacy

Xavier Aldana Reyes

At first sight, Barcelona seems an unlikely place for Gothic mysteries. As a balmy Mediterranean city averaging temperatures in the 20–24 °C band from June to August (*Holiday-Weather.com*, 2018), its marketing campaigns have consistently aimed to project an image of joyful, family-friendly sun and sea holidays.[1] Spain is well-known for its food, so the cuisine (especially seafood and fish) is naturally another highlight, as are sports (thanks to the connections to Football Club Barcelona), shopping (haute couture shops in Passeig de Gràcia) and culture (especially the Gaudí modernist architecture trail that includes la Pedrera, la Casa Batlló and Parc Güell). This means Barcelona tries to appeal to the connoisseur in search of a distinctive spring or summer break, and that it can capitalise

The original version of this chapter was revised: The incorrect author's given name and family name in chapter title page and running head have been corrected. The corrections to this chapter are available at https://doi.org/10.1007/978-3-030-43777-0_15

X. Aldana Reyes (✉)
Manchester Metropolitan University, Manchester, UK
e-mail: X.Aldana-Reyes@mmu.ac.uk

H.-G. Millette and R. Heholt (eds.),
The New Urban Gothic, Palgrave Gothic,
https://doi.org/10.1007/978-3-030-43777-0_14

on both party-hungry tourists and those after a more sedate and introspective experience. It has largely succeeded in creating this perception, too. Writing for *The Telegraph* in 2017, journalist Sally Davies recommended Barcelona on the grounds that it 'combines everything that is most charming about Mediterranean cities – a relaxed pace, months of endless sunshine, unbeatable food – with the cultural and design clout of almost any city in the cold north', and added that it is 'a city with a proud sense of identity'. Barcelona also placed tenth in Forbes's top 10 most visited cities in the world in 2018 ('The World's Top 10…', 2018) and was ranked sixth in the top 10 'Most Popular cities in Europe' in 2017 by *The Huffington Post* ('These Are the Ten…', 2017). While the turmoil over independence for Catalonia may have somewhat affected its peaceful and idyllic image, Barcelona still seems a far cry from the urban horrors that populated the Victorian novel (and London, in particular) and which led to what Robert Mighall has called a 'Gothic *of* the city' (1999, p. 30, italics in original). Even in Spain's own literary history, Barcelona was, until recently, mostly pictured as an industrialist dreamworld of opportunity, as in Eduardo Mendoza's classic novel *La ciudad de los prodigios* (1986) [The City of Marvels].

And yet, this is not the whole picture. Barcelona has been Gothicised in literature and film in the twenty-first century in the novels of Carlos Ruiz Zafón, as well as in the various literary projects connected to the case of serial killer Enriqueta Martí, the 'vampire of el Raval'—including the novels *El misterio de la calle poniente* (2007) [The Mystery of Poniente Street], by Fernando Gómez, Marc Pastor's *La mala dona* (2008) [The Bad Woman], Elsa Plaza's *El cielo bajo los pies* (2009) [The Sky under Our Feet] and Miguel Ángel Parra, Iván Ledesma and Jandro González's graphic novel *La vampira de Barcelona* (2017) [The Vampiress of Barcelona].[2] These texts point towards another 'hidden Barcelona', one that has either been forgotten or is not widely talked about. Crucially, the city's current tourist industry is also premised on artificiality. Few visitors may realise that Barcelona's beachfront and marina are a recent development that followed a concerted effort to restructure the city in preparation for the defining Olympic Games of 1992. These involved a significant financial investment in infrastructure, and transformed Barcelona radically, making it appealing as a destination and rescuing it from decades of governmental neglect under Franco (Taylor 2012). Visitors may also be unaware that the only self-professedly Gothic spot, el Barri Gòtic, or Gothic quarter—a melange of Roman, medieval, nineteenth- and twentieth-century architecture—was largely constructed between 1927 and 1970 to create an attractive, national and tourist-friendly destination. As art historian Agustín Cócola Gant suggests, the Barri Gòtic is less a real historical place than a good example of how Barcelona's 'heritage [has been] exploited as a channel for promoting the city' (2014, p. 30). In terms of Barcelona-associated literature, the popularity of Eugène Sue's *Les Mystères de Paris* (1843) [The Mysteries of Paris] gave rise to a number of national imitations in the nineteenth century that saw, among others, the publication of Antonio Altadill's *Barcelona y sus misterios* (1860) [Barcelona and Its Mysteries], reissued in Catalan in 2015. The novel, adapted to cinema in 1915 as an eight-part serial, explored the criminal underbelly of the city, an aspect that would be resurrected in the twenty-first century.

This chapter seeks to both position the self-aware recuperation of a darker Gothic Barcelona in the novels of Ruiz Zafón's series, *The Cemetery of Forgotten Books* (2001–16), and to legitimise its potential value as deliberate artificial construct. While it is possible to read the Gothicisation of Barcelona as a contrived marketing exercise interested in exploiting its central locations, it is as creative an act as that of the consolidation of a 'bright' Barcelona in the popular imagination. While the latter strives to promote a profitable tourist industry, the former seems concerned with rethinking the city's history, especially the legacy of the Civil War on the present. As such, Barcelona does not become Gothic; it is inherently so: the process of unearthing its dark side can be read as a political statement.

Ruiz Zafón's Gothic Barcelona

Ruiz Zafón's preference for the supernatural register and the Gothic mode of writing was already apparent in his early Young Adult (YA) fiction and, especially, in his 1999 novel *Marina* (published in English in 2013), set in Barcelona and the Sarrià neighbourhood, where he grew up (see Aldana Reyes 2017, pp. 172–176). From a cemetery that does not show up in maps to a subterranean room full of creepy dolls, to the Gothic quarter, the sewers and even the ruins of the Gran Teatro Real [Grand Royal Theatre], the locations used as the backdrop for the mysteries in the novel evoke secrecy, decay, faded grandeur, death and, importantly, the past. The impossibility of returning to better times, as well as the immense difficulty of fully expunging painful memories and overcoming their legacy, are pervasive motifs in Ruiz Zafón's work, both in his YA and adult fiction. The impervious nature of the past is also imbued with a strong sense of place. Our personal histories, Ruiz Zafón suggests, cannot be extricated from our sociohistorical contexts, from the places where we grew up and the people we have known. This is the case for the characters of the four-volume *The Cemetery of Forgotten Books* series, made up of the novels *La sombra del viento* (2001), published in English as *The Shadow of the Wind* (2004), *El juego del ángel* (2008), published in English as *The Angel's Game* (2009), *El prisionero del cielo* (2011), published in English as *The Prisoner of Heaven* (2013), and *El laberinto de los espíritus*, published in English as *The Labyrinth of Spirits* (2018). Of these books, which may be read independently but contain interconnected stories, by far the most popular is *The Shadow of the Wind*. In it, a young boy is taken to the titular library, hidden in the heart of Barcelona, by his dad. There, first-timers must select a book from its shelves and promise to protect it with their life. The boy's choice, a book by one Julián Carax, sets off a long and complex series of events that will end in the apparent return of the dead writer and a hunt for the last remaining copy of the cursed book. Set in 1950s post-war Spain, the novel established the basic parameters for the other volumes in the series, all of which are invariably set during or after the Civil War in Barcelona (also Madrid in *The Labyrinth of Spirits*) and follow the adventures of characters connected to the boy's family.

Echoes of the Victorian Gothic, such as the creations of Mary Shelley, Robert Louis Stevenson and Gaston Leroux, are perceptible in Ruiz Zafón's writing. Sometimes the debt to other writers is made explicit in direct allusions, for example in the references to Alexandre Dumas and his *Le Comte de Monte-Cristo* (1844–5) [The Count of Monte Cristo]. As an exponent of an eminently transnational and glocal (both global and local) form of the Gothic, Ruiz Zafón borrows from external sources (the British and French Gothic traditions) and hybridises them with Spanish pulp literature (the nineteenth-century 'folletín', or serial, novel features prominently in his work, especially in *The Angel's Game*). In so doing, he does not just recycle, he also innovates. This means his literature is both personal and eminently exportable—its literary influences make it more universal. Attempting to capture the individuality of Ruiz Zafóns' Gothic brand, journalist Sergio Vila-Sanjuán proposes that the main characters in his novels tend to be 'damned writers with burnt faces, lonely and tormented teenagers, devoted and altruistic heroines, twisty-turny alleys and evil destinations' (2008, p. 38), but, above all, it is 'a dark Barcelona, violent, melodramatic, frightful, with red skies and blood lakes under its supposedly calm streets' (p. 38, my translation) that stands out as the protagonist.[3] The most significant trait in Ruiz Zafón's fiction, then, is the development of a strand of the Gothic that is eminently urban and, following *Marina*, largely confined to the city of Barcelona, its heritage and modern history. This is important because his adult novels have also left behind the magic of his YA fiction in favour of a form of the explained supernatural. Supernatural phenomena in the four volumes that constitute *The Cemetery of Forgotten Books* are always, in fact, proven to be manipulative ruses or else the hallucinations of madmen. Despite an interest in the Gothic in all its manifestations, there is a significant difference between Ruiz Zafón's early writing and his later adult novels. *Marina* sits as a bridge novel in his oeuvre, still eminently fantastic but already set in Barcelona, and marked by the beginnings of an interest in social realism that would be much more fully articulated in *The Cemetery of Forgotten Books* series.

The Shadow of the Wind is set in Barcelona, a very dank and rainy one—an extreme example of the use of pathetic fallacy in fiction. The city described in the opening paragraph, when protagonist Daniel Sempere recalls his first visit to the library in a summer day in 1945, is one 'trapped beneath ashen skies as dawn pour[s] over Ramblas de Santa Mónica in a wreath of liquid copper' (Ruiz Zafón 2005, p. 1). This atmosphere of gloom and doom rarely lifts, and plays an important role at the level of mood, reflecting the air of mystery and dread that pervades the novel. Barcelona is, in fact, recalled by another character as a 'city of shadows' (p. 25). Later, Barcelonan streets near the central Calle Canuda appear to be full of 'faceless people escaping from offices and shops' as they are swept by 'an icy wind' (p. 177); 'the corner of Gran Vía and Paseo de Gracia [is] under a leaden sky that st[eals] the light of day' (p. 233); Calle Balmes 'fade[s] behind a curtain of liquid velvet' (p. 236) and Plaza Santa Ana is equally visited by 'heavy rain' (p. 353). Memories of the past do not paint a more comforting picture. Jacinta, a former governess who helps Daniel later in the novel, remembers the Barcelona of Carax's younger years as an industrial, corrupted hell, as 'a sinister, hostile city, full of clogged mansions, full of factories that poured forth, poisoning the air with coal and sulphur' (p. 270)—a 'dark city' (p. 271). Another character describes the Barcelona of his honeymoon as 'a little world of fog' (p. 406). 'This city is a sorceress [...]. It gets under your skin and steals your soul without you knowing it' (p. 496), claims Daniel's friend, Fermín, in the novel's closing pages. The result of all this conscious evocation of rain, darkness and hostility is indeed a spellbinding and drawn-out process of atmospheric priming that does more than create a suitable stage for the novel's many Gothic motifs. Barcelona becomes its own Gothic persona, scarred by the past, ashen and incapable, it seems, of overcoming its traumatic history.

As Glennis Byron has noted, the process of Gothicisation of Ruiz Zafón's work and his Barcelona needs to be understood within an industry context in which publishers gradually accentuated the Gothic elements in his novels when marketing them. *The Shadow of the Wind* was not originally sold as Gothic fiction in Spain or the UK, but as a literary thriller. It was only later, after critics like best-selling horror writer Stephen King declared the novel a Gothic masterpiece, that his work began to be more actively promoted in this way (Byron 2012, p. 76). In an even more interesting turn of events, the still active '*Shadow of the Wind* Walking Tour' soon developed to take readers into the heart of the novel's events and thus into a fabricated Gothicised version of Barcelona that includes the claustrophobic streets of the Gothic quarter.[4] As Byron (2012, p. 78) points out, however, anyone looking for the gloomy Barcelona of *The Shadow of the Wind* in this walk will be inevitably disappointed by the reality of the bustling, sunny, Catalan metropolis. Neither is the apparent excuse that Ruiz Zafón's Barcelona strives to be 'a purely literary [...] representation of the city' (Ruiz Zafón, quoted in Porter 2008), as the author claimed in at least one interview, entirely satisfying. The decision to include promotional maps in subsequent editions of the book, although perhaps not an authorial choice, is rather telling. Naturally, place writing is symbiotic: connections to a city and its tourist industry can boost book sales, and there is a lot to be said for the way in which *The Shadow of the Wind* helped put Barcelona and Spanish literature on the cultural map in an unprecedented way. Far from intending to criticise the pecuniary interests behind such moves, I want to suggest that there might be an associated or alternative rationale, more artistically significant, for the Gothicisation of Barcelona in *The Cemetery of Forgotten Books* series. The Barcelona in these novels is not a contemporary one, but very specifically a Civil War and post-war version of it. As I show, the political and national significance of this cannot be readily dismissed, even though it appears to have gone unnoticed by readers of the novels who are not attuned to Catalonia's relationship with central(ist) Spain and Franco's ideological legacy. The Barcelona in Ruiz Zafón is Gothic because it takes place during what is perceived as an oppressive and barbaric past that, like a family curse, has continued to return to the lives of modern Catalans. The epigraph in *The Prisoner of Heaven*, taken from Carax's fictional book of the same title, indicates as much when it calls Barcelona 'the city of the damned', a place 'trapped in ashes and silence' (Ruiz Zafón 2013, n.p.).

Barcelona and Franco's Gothic Legacy

The Gothicisation of Barcelona in Ruiz Zafón comes hand in hand with a repudiation of the crimes committed during the Civil War. It is not a coincidence that Carax's novel, which shares its title with Ruiz Zafón's, is published only a couple of weeks before the onset of the Civil War in 1936, and is thus destined for oblivion. As is revealed in *The Shadow of the Wind*, Carax is not an evil Gothic villain, but a tragic victim for whom Daniel will ultimately risk his own life. The real antagonist is inspector Fumero, a mercenary policeman who tries to kill Carax for kissing Penélope, a girl with whom he also fell in love. Although clearly a traumatised man, Fumero is presented unsympathetically. Two of the few things known about his past is that he used to take pleasure in torturing animals and that he shot and killed his own mother while still a child. When, after being beaten up by Fumero, Fermín talks about the inspector, the latter explains that Fumero is a man who has 'no fear and no scruples', no sense of loyalty or beliefs (Ruiz Zafón 2005, p. 332). Although in *The Shadow of the Wind* the focus is Fumero's own moral depravity (he works for anarchists and communists before joining fascist forces), the connections between him, Franco's dictatorship and the war grow tighter as the series advances. In *The Shadow of the Wind* he becomes Daniel's direct antagonist: while Daniel's mission involves uncovering the buried history of Carax and his deceased beloved, Fumero stands as a reactionary gatekeeper. He despises 'people who stir up shit from the past!' (p. 294) and insists that '[t]he past must be left alone' (ibid.). It is clear that this message does not resonate with Ruiz Zafón's artistic agenda. Even if the engagement with the Francoist past is not as openly critical and candid here as it is in *The Prisoner of Heaven* or *The Labyrinth of Spirits*, for Ruiz Zafón to forget is to let others rule the present; to dig for it, on the contrary, is cathartic. This passage, which makes direct reference to the Civil War, is telling:

> Nothing feeds forgetfulness better than war [...]. We all remain silent and they try to convince us that what we've seen, what we've done, what we've learned about ourselves and about others, is an illusion, a nightmare that will pass. Wars have no memory, and nobody has the courage to

> understand them until there are no voices left to tell what really happened, until the moment comes when we no longer recognize them and they return, with another face and another name, to devour everything they left behind. (p. 443)

In Ruiz Zafón's interesting rethinking of the Gothic tradition, the past is both something that needs to be protected from the ravages of the present (symbolized by the precious last copy of Carax's literary masterpiece) and feared, because, if unacknowledged, memories will return to haunt and destroy us. It is pertinent that Barcelona should be the backdrop for a story seeking to treat national history (conflated here with the personal past) in a Gothic manner. If the Carlist Wars were at one point the most Gothicised historical event in popular Spanish literature, the silence still surrounding the Civil War and Franco's dictatorship makes it perfect fodder for conspiratorial mysteries and the brutal unleashing of repressed behaviours.[5]

The novel's stance, one that begins by foregrounding the need to put an end to the silence around the Civil War and by suggesting that the Spanish take ownership both individually and collectively for the mistakes of the past, resonates with the country's own gradual political and social awakening to its modern history. 'El pacto del olvido' [The Pact of Forgetting], drawn after Franco's death in 1975, aimed to smooth the transition from a dictatorship to democracy by outlawing the persecution of those who had committed crimes against humanity during the war. It only began to be actively challenged in the twenty-first century. The year 2000 saw the foundation of the Association for the Recovery of Historical Memory, whose members started collecting oral and written testimonies of the victims of the regime, and in 2004 the socialist government proposed a ministerial commission to 'restituir la dignidad y la memoria' [restore the dignity and memory] of the war's casualties, as well as to give rights to their descendants. In 2007, the Historical Memory Law was passed by the Congress of Deputies. It allowed for, among other things, the state supervision of grave searches in what is hard not to conceive of as a literal digging up of the war's dead. And it is not just Republicans whose slumber was disturbed. In 2017, Spanish lawmakers called for Franco's tomb to be removed from the basilica in the 'Valle de los Caídos' [Valley of the Fallen], a state-funded monument commemorating the victims of the war which had, ironically, become a pilgrimage site for far-right groups. The exhumation of Franco's remains was approved in

August 2018 and carried out in 2019. Antonio Pradas, a deputy from the socialist party, has complained that there are victims of the war who got 'no recognition', and that one 'can only turn the page of history when all victims are recognised. The way to end the divisions in Spain is not by covering up and shutting up' (Pradas, quoted in Buck 2017). Lawmakers have also urged the government to create a 'truth commission' to investigate the crimes committed during Francoism, a timely initiative given that the United Nations have repeatedly urged Spain to repeal the 1977 Amnesty Law that gives the Pact of Forgetting a legal basis. Unsurprisingly, both the Church and the ruling Popular Party resisted this motion. This is not necessarily out of character. At least one journalist noted in 2017 how, since they came into power in 2011, the Popular Party had not removed a single of the 135 Francoist monuments they were tasked with retiring as part of the country's compliance with the Historical Law Memory (Díez 2017). By contrast, the Socialist Party retired nearly 600 of said monuments between 2009 and 2011.

It was also around the turn of the century that what has been termed the 'memory boom', or the publication of a substantial number of fictional texts and memoirs chronicling the war (Richards 2013, p. xii), really took off. Films and novels such as *Libertarias* (Vicente Aranda, 1996) [Freedomfighters], *La lengua de las mariposas* (José Luis Cuerda, 1999) [Butterfly's Tongue] or *El lápiz del carpintero* (Antón Reixa, 2003) [The Carpenter's Pencil], and the television series *Cuéntame cómo pasó* (TVE1, 2001–present) [Tell Me How It Happened] were not the first to break the silence, but they signalled a definite move towards a consistent questioning of the systemic violence of the repressed past. Although the preferred mode through which to explore the Civil War has tended to be the most scrupulous of social realisms, a significant and growing number of Gothic treatments have surfaced in the twenty-first century. 2001, the same year *The Shadow of the Wind* appeared in Spain, also saw the release of Guillermo del Toro's celebrated Mexican–Spanish co-production *El espinazo del diablo* (2001) [The Devil's Backbone]. Del Toro's tale of a haunted small home for orphans in a remote expanse of land crowned by a defused bomb did much to establish the tropes of the Civil War Gothic: the return of unresolved trauma, especially of that caused by brutal and unnecessary violence, and generally manifesting in the form of a supernatural being; the legacy of the past on the present, which it inhabits and haunts; the portrayal of Francoist forces as heartless, licentious and sadistic; and the use of wartime and early post-war settings as Gothic places associated with the dark, death and even evil. *El laberinto del fauno* (2006) [Pan's Labyrinth], del Toro's second film about the Civil War, did much to galvanise this Gothic imagery and trappings via magical realism and fairy-tale lore. *Insensibles* (2012) [Painless], a Spanish co-production with France and Portugal set in an asylum in Canfranc (Aragón Valley), is a film that proposes the victims of the war were not just the dead, but also the generations that followed. In it, an adopted man in need of a marrow transplant discovers his real father was one of many children suffering from an exceptional medical condition which rendered them unable to experience pain. Benigno (Tomás Lemarquis), renamed Berkano (the rune name for 'becoming'), was one of the smartest and, it is revealed, eventually became one of the Francoist regime's most prolific torturers. The film's conceit needs little decoding: the generation raised under Franco are unable to feel and express the trauma of the war, and thus have become active ideological participants in its reification as unprocessed memory.

Like these other Civil War Gothics, Ruiz Zafón's books interweave formulae from Gothic literature with a revisionist social realism strongly imbued with the need to recuperate the Civil War as a traumatic event the perception of which has been affected by recent historical and social changes. The horror in the novels is increasingly derived from the war, from the subsequent fascist regime and from its followers. One of the few truly fear-inducing scenes in *The Prisoner of Heaven* involves a nightmare where a captured Fermín witnesses 'a bottomless mass grave strewn with corpses [...], the flood of ghostly bodies stirring like an eddy of eels. The dead bodies opened their eyes and climbed the walls, following him' (Ruiz Zafón 2013, p. 151). Similarly, *The Labyrinth of Spirits* opens with a chapter set in Barcelona's harbour during the 1938 nationalist bombings that reduced large parts of the city to rubble. The bottom of the sea is described as seething with 'dozens of handcuffed corpses' [decenas de cadáveres esposados] that form a macabre 'submarine cemetery' [cementerio submarino] (Ruiz Zafón 2016, p. 51, my translation). The Civil War is imagined in all its cruelty and viscerality, Gothicised as much as the setting. Even the library that becomes the catalyst of all events in the series is appropriately named. The Cemetery of Forgotten Books stands as a poetic image for those killed and forgotten, an 'endless necropolis' where volumes 'remain unexplored, forgotten forever' (Ruiz Zafón 2005, pp. 74–75) until rescued by a daring reader who may become their keeper. Similarly, the spirits in the labyrinth of Ruiz Zafón's latest and last instalment epitomise the predicament of the younger Spanish generations, who appear unmarked by the war, yet find themselves locked in a traumatic return marred by the very desire to avoid or repress painful memories.

Jo Labanyi (2007) has suggested that the traumatic paradigm popular in Holocaust Studies may not be appropriate for the case of the Spanish Civil War and that the reason for the relative lateness of the memory boom may be due to a reluctance on the part of the interlocutors to tell their stories. She has also argued that the decision to break with the past after Franco's death may have more to do with not wanting to let the past affect the future than with its deliberate silencing. This is true, but only insofar as intention is concerned. As the controversial 2017 vote for Catalan independence demonstrated, the very idea that the past should not affect the future is problematic, for it masks a desire to let the status quo go unchallenged. Following the police attacks on Catalan independence voters, widely reported by news programmes around the world, protesters (and eventually Parliament ex-president Núria de Gispert) appropriated a new version of the popular phrase 'esto con Franco no pasaba' [this would have never have happened under Franco], used to express bewilderment and disgruntlement at apparently lax moral standards. 'Esto con Franco sí pasaba' [this was already happening under Franco] appeared in Tweets and was chanted in many of the demonstrations covered by the media ('El grito...', 2017). This phrase signals that, for some independentists, there is a clear connection between the oppressive rhetoric of Mariano Rajoy, Prime Minister of Spain from 2011 and 2018, and that of Franco. Even before the recent contentious vote, Catalans had been confused by moves to renew the memory of the dictator. For example, in 2016, there was an attempt to resurrect a headless statue of Franco in the Born market as part of a commemorative exhibition which Barcelona's deputy mayor claimed aimed to 'denounce the [era's] crimes' (Agence France-Presse, 2016). This decision met with the disapproval of nationalists, and received a good pelting of eggs. Catalonia, and Barcelona as its capital, is a part of Spain with a long history of resistance to the centralist policies best encapsulated by Franco's idea of a unified nation at the expense of regional exceptionalism. Its Gothicisation in Ruiz Zafón's work is not apposite, but political.

One of the characters in Javier Cercas's *El monarca de las sombras* (2017), published in English as *Lord of All the Dead* in 2019, a biographical novel about a nineteen-year-old Falangist who died during the Civil

War, establishes at one point the need to accept the familial and personal past as a necessary step towards the communal negotiation of a country's past. Rather than 'hide it', the novel suggests, we need to 'accept it as it is, once and for all, in all its toughness and complexity, without sweetening or embellishing or under rug sweeping it' (Cercas, 2017, p. 45, my translation). The Civil War remains Spain's favourite Gothic tale because the country has still not openly dealt with the legacy of the war, with its impact on the lives of those directly affected by it. It is difficult for this to take place, given that the crimes committed during Francoism have not even been legally recognised, let alone sanctioned. Whether the past can ever be truly laid to rest is a complex issue, but recognition of its effects on the present and the way that current discourses around nationalism and patriotism have been coloured by previous acts of violence (physical and ideological) is necessary if we are to see the end of the type of alienation, fear and anger that Ruiz Zafón explores in his fiction. In this context, Barcelona becomes a Gothic microcosm, a playground where heavy and artificial symbolism is forced upon the city in order to conjure up 'its reverse, a damned reflection' (Ruiz Zafón 2016, p. 266, my translation) [su reverso, un reflejo maldito]. The simultaneous retrojection of this dark Barcelona to the days of the Civil War and its aftermath in the novels of *The Cemetery of Forgotten Books* intrinsically connects both ideas: the barbaric setting echoes the oppressive regime; its forgotten victims are rendered ghostly Caraxes in need of literal and metaphorical rescue. Only by coming to terms with the legacy of the past, like Daniel does in *The Shadow of the Wind*, can Spain move into the future. The Gothic Barcelona in Ruiz Zafón, one of its many iterations, is therefore not merely a literary representation, but a tragic mirror where past and present collapse.

Notes

1. For Barcelona's own promotion of its public image, see, among many others, *ConocerBarcelona.com* (2018).
2. El Raval is a neighbourhood in central Barcelona.
3. In the original, 'escritores malditos de rostro quemado, adolescentes solitarios y atormentados, heroínas entregadas y altruístas, retorcidos callejones y destinos maléficos' and 'una Barcelona oscura, violenta, melodramática, tremenda, con cielos rojos y lagos de sangre bajo sus supuestamente apacibles calles'.

4. The route can be downloaded as a PDF from the author's website (as of 12 February 2018), although official tours may also be booked via external websites.
5. The Carlist Wars were a series of nineteenth-century conflicts spurred on by the succession to the throne, following the death of Ferdinand VII of Spain, of Queen Regent Maria Cristina. The country was split between Cristinos, those who supported Maria Cristina and were more liberal, and those who supported Carlos V, pretender and brother of Ferdinand. The wars had a strong regional inflection, affecting the Basque Country and Catalonia.

References

Agence France-Presse. 2016. Franco Statue Pelted with Eggs as Polarising Exhibition Opens in Barcelona. *The Guardian*, 18 October. Available at: https://www.theguardian.com/world/2016/oct/18/franco-statue-pelted-with-eggs-as-polarising-exhibition-opens-in-barcelona. Accessed 5 June 2017.

Aldana Reyes, Xavier. 2017. *Spanish Gothic: National Identity, Collaboration and Cultural Adaptation*. Basingstoke: Palgrave.

Buck, Tobias. 2017. Spanish Lawmakers Call for Franco Tomb to Be Removed. *Financial Times*, 9 March. Available at: https://www.ft.com/content/23773dce-04a3-11e7-ace0-1ce02ef0def9. Accessed 5 June 2017.

Byron, Glennis. 2012. Gothic, Grabbit and Run: Carlos Ruiz Zafón and the Gothic Marketplace. In *The Gothic in Contemporary Literature and Popular Culture: Pop Goth*, ed. Justin D. Edwards and Agnieszka Soltysik Monnet, 71–83. Abingdon and New York: Routledge.

Cercas, Javier. 2017. *El monarca de las sombras*. Barcelona: Literatura Random House.

Cócola Gant, Agustín. 2014. The Invention of the Barcelona Gothic Quarter. *Journal of Heritage Tourism* 9 (1): 18–34.

Davies, Sally. 2017. The World's 20 Most Visited Cities—Where Does London Rank? *The Telegraph*, 22 May. Available at: http://www.telegraph.co.uk/travel/galleries/the-worlds-most-visited-cities-where-does-london-rank/barcelona/. Accessed 12 February 2018.

Díez, Anabel. 2017. El Gobierno del PP no ha retirado ni un solo vestigio franquista en sus cinco años de mandato. *El País*, 10 March. Available at: https://politica.elpais.com/politica/2017/03/10/actualidad/1489133313_429831.html. Accessed 5 June 2017.

Labany, Jo. 2007. Memory and Modernity in Democratic Spain: The Difficulty of Coming to Terms with the Spanish Civil War. *Poetics Today* 28 (1): 89–116.

Mighall, Robert. 1999. *A Geography of Victorian Gothic: Mapping History's Nightmares*. Oxford: Oxford University Press.

Porter, Steve. 2008. Books Hold No Passport – Carlos Ruiz Zafón Discusses *The Shadow of the Wind*. *Three Monkeys Online*, October. Available at: http://www.threemonkeysonline.com/books-hold-no-passport-carlos-ruiz-zafn-discusses-the-shadow-of-the-wind/. Accessed 12 February 2018.

Richards, Michael. 2013. *After the Civil War: Making Memory and Re-making Spain Since 1936*. Cambridge: Cambridge University Press.

Ruiz Zafón, Carlos. 2005. *The Shadow of the Wind*, trans. Lucia Greaves. London: Weidenfeld and Nicholson.

———. 2013. *The Prisoner of Heaven*, trans. Lucia Graves. London: Phoenix.

———. 2016. *El laberinto de los espíritus*. Barcelona: Planeta.

Taylor, Adam. 2012. How the Olympic Games Changed Barcelona Forever. *BusinessInsider.com*, 26 July. Available at: http://www.businessinsider.com/how-the-olympic-games-changed-barcelona-forever-2012-7?IR=T. Accessed 12 February 2018.

Vila-Sanjuán, Sergio. 2008. El mundo gótico de Carlos Ruiz Zafón. In *Barcelona Gothic*, ed. Carlos Ruiz Zafón, 38–41. Barcelona: La Vanguardia.

Websites

Barcelona: Annual Weather Averages. 2018. *Holiday-Weather.com*. Available at: http://www.holiday-weather.com/barcelona/averages/. Accessed 12 February 2018.

ConocerBarcelona.com. 2018. Available at: https://www.conocerbarcelona.com/. Accessed 12 February 2017.

El grito de "esto con Franco sí pasaba" se viraliza. 2017. *ElPlural.com*, 20 September. Available at: https://www.elplural.com/comunicacion/2017/09/20/el-grito-de-esto-con-franco-si-pasaba-se-viraliza. Accessed 5 November 2017.

The World's Top 10 Most Visited Cities: 10. Barcelona. 2018. *Forbes.com*. Available at: https://www.forbes.com/pictures/ehlk45iehm/10-barcelona/#18b3bcdd606f. Accessed 12 February 2018.

These Are the 10 Most Popular Cities in Europe (According to You). 2017. *Huffington Post*, 6 December. Available at: https://www.huffingtonpost.com/hoppercom/these-are-the-10-most-pop_b_7207258.html. Accessed 12 February 2018.

Correction to: A Gothic Barcelona?: Carlos Ruiz Zafón's *The Cemetery of Forgotten Books* Series and Franco's Legacy

Xavier Aldana Reyes

Correction to:
Chapter 14 in: H.-G. Millette and R. Heholt (eds.),
***The New Urban Gothic*, Palgrave Gothic,**
https://doi.org/10.1007/978-3-030-43777-0_14

The original version of this chapter was inadvertently published with incorrect author's given name and family name in chapter title page and running head, which have now been corrected. The corrections to the chapter have been updated with the changes.

The updated version of this chapter can be found at
https://doi.org/10.1007/978-3-030-43777-0_14

H.-G. Millette and R. Heholt (eds.),
The New Urban Gothic, Palgrave Gothic,
https://doi.org/10.1007/978-3-030-43777-0_15

Index

H.-G. Millette and R. Heholt (eds.),
The New Urban Gothic, Palgrave Gothic,
https://doi.org/10.1007/978-3-030-43777-0

The manufacturer's authorised representative in the EU is Springer Nature Customer Service Centre GmbH, Europaplatz 3, 69115 Heidelberg, Germany. If you have any concerns regarding our products, please contact ProductSafety@springernature.com

Printed and bound by CPI Group (UK) Ltd, Croydon, CR0 4YY

30/09/2026

02225646-0002